A FACTS ON FILE PUBLICATION

CUBA, THE U.S. & RUSSIA
1960-63

A journalistic narrative of events in Cuba and of Cuban relations with the U.S. and the Soviet Union

Compiled by the Editors of
Facts on File and News Year

Lester A. Sobel, Editor-in-Chief

INTERIM
HISTORY

FACTS ON FILE, INC.
119 West 57th Street, New York, N.Y. 10019

CUBA, THE U.S. & RUSSIA 1960-63

© Copyright, 1964, by Facts on File, Inc.

Library of Congress Catalog Card No. 64-55088
ISBN 0-87196-150-4

9 8 7 6

PRINTED IN THE UNITED STATES OF AMERICA

Contents

1963

Cuba, the U.S. & Russia 1960-63

The years 1960-63 were marked by a series of incidents that underlined the growing antagonism between the U.S. and the Fidel Castro régime of Cuba. Many events of this period added to the strain in U.S. relations with Russia over Castro's decision to form ever-closer ties with the Soviet bloc.

The U.S. severence of diplomatic relations with Cuba in 1961 was followed by a tragically unsuccessful attempted invasion of Cuba by U.S.-assisted Cuban exiles. A tense confrontation between the U.S. and the USSR took place in 1962 with the discovery that the Soviets had built a network of "aggressive" missile bases in Cuba. The threat of a U.S.-Soviet war over the issue dissipated only after Russia yielded and agreed to dismantle the bases.

This book tells in detail the story of the events mentioned above as well as of other occurrences involving Cuba, the U.S. and Russia during the 4-year period 1960-63. Except for some editorial revision, largely minor, the material of this book comes directly from the pages of the 4 volumes of News Year that cover the period involved.

Antagonisms flared during 1960 over Cuba's seizures of U.S. property, over Cuban Premier Castro's mounting collaboration with the Soviet bloc and over mutual U.S.-Cuban charges of aggressive intent. Hemisphere foreign ministers met in San José and approved an anti-Cuban declaration condemning Communist encroachment in the Americas.

U.S.-Cuban Tensions

Seizures Protested. U.S. Amb. Philip W. Bonsal delivered in Havana Jan. 11 a note protesting the Cuban government's seizure of property owned by U.S. citizens. The note was rejected by the Cuban government the same night. The note charged that the confiscation & removal of equipment, seizure of cattle, cutting & removal of timber and the plowing under of pastures had been carried out "without court orders and frequently without any written authorization." Whereas U.S. notes of June 11 & Oct. 12, 1959 had expressed full support of "soundly conceived programs" of land reform, the new note asserted that such programs were not furthered "by the failure of . . . Cuba to recognize the legal rights of U.S. citizens who have made investments in Cuba."

Accusations Traded. Attacks on the U.S. & Spain by Castro in a series of speeches Jan. 18-21 provoked (1) recall to Washington of the U.S. ambassador, (2) expulsion from Cuba of the Spanish ambassador and (3) a statement by Pres. Eisenhower defending U.S. policy but promising "no reprisals" for Castro's actions.

In a radio speech Jan. 18, Castro charged that the U.S. had violated diplomatic custom by making public, before the Cuban government received it, the Jan. 11 note against Cuba's method of confiscating American-owned property. Castro declared that his revolution had released Cuba from decades of foreign domination—first by Spain, then by the U.S.

During a long TV interview in Havana the night of Jan. 19-20, Castro charged that U.S. Amb. Bonsal & Spanish Amb. Juan Pablo de Lojendio, Marquis of Velisca, had conspired with counter-revolutionaries against the Castro regime. Castro denounced talk in the U.S. of amending the 1934 Sugar Act under which the U.S. assured Cuba 2¢-3¢ more a pound for sugar than it would bring on the world market. Castro said the Sugar Act was designed primarily "to protect American cane interests," not to subsidize sugar for the benefit of Cuba. Spanish Amb. Lojendio, enraged by Castro's remarks, went to the TV studio and broke into the program to protest that he had been "slandered & vilified." Castro replied by telling Lojendio, "You have 24 hours to get out of Cuba"—an expulsion notice seconded immediately by Cuban Pres. Osvaldo Dorticós Torado.

Bonsal, recalled to Washington Jan. 21 for consultation, reported to State Secy. Christian Herter Jan. 24 and at the White House Jan. 25.

Pres. Eisenhower then issued at his news conference Jan. 26 a policy statement deploring Castro's "unwarranted attacks" but promising that the U.S. was "not going to be a party to reprisals or anything of that kind. . . . Certainly we are not going to intervene" in Cuba's "internal affairs."

Mr. Eisenhower said: The Castro regime had sought "to create the illusion of aggressive acts & conspiratorial activities aimed at the Cuban government & attributed to U.S. officials or agencies"; actually, the U.S. government's record in curbing leaflet, gun-running & other raids on Cuba from American territory "compares very favorably with that of Cuba, from whose territory a number of invasions directed against other countries have departed during the past year"; the U.S. had not protested Cuba's confiscation of property owned by Americans but had objected to seizure without official inventory or receipt and to the proposed compensation by long-term bonds.

Mr. Eisenhower said the U.S. recognized Cuba's right to undertake "social, economic & political reforms" as it saw fit but would "continue to bring to the attention of the Cuban government any instances in which the rights of [U.S.] citizens, under Cuban law & under international law, have been disregarded and in which redress under Cuban law is apparently unavailable or denied." He said the U.S. wanted to resolve disputes through "diplomatic negotiations" or other "international procedures."

Cuban Pres. Dorticós responded with a statement Jan. 27 saying Cuba "laments the progressive deterioration of relations," but he blamed it on American "misunderstanding" of the Castro regime's ideals & goals. He agreed that disputes should be resolved through diplomatic negotiation. He promised "no confiscation of foreign property, but expropriation compensated in the periods & manner which our constitution authorizes & our financial situation permits."

Bonsal returned to Havana Mar. 20 and delivered 3 State Department notes replying to 3 Cuban notes.

The first U.S. note said: There had been increased tensions in the Caribbean since the establishment of the Castro revolutionary government; hopes that the new government would reestablish peace & tranquility in the area had been dashed when it had requested "armaments far in excess of any conceivable need for self-defense"; this was the reason for the U.S. Commerce Department's refusal to issue licenses for the export of helicopters to Cuba.

The 2d U.S. note said Cuban labor leader Federico Figueras Larrazabal had been fired from a job at the U.S. naval base at Guantánamo because he had made repeated slanderous statements about the U.S. It denied that "many" Cuban army members had

---

C.I.A. ON CASTRO. Gen. C. P. Cabell, deputy director of the U.S. Central Intelligence Agency, in testimony Nov. 5, 1959 to the Senate Internal Security Subcommittee made public Jan. 30, gave this CIA estimate of Cuban Premier Fidel Castro:

He was "not a Communist" but "certainly not anti-Communist." His "extreme policies . . . are being exploited by the Communists to the maximum extent." He had "delegated authority in key areas to persons known to be pro-Communists or who are susceptible to exploitation by Communists. In turn, he appears to be increasingly susceptible to Communist propaganda." The CP "viewpoint is that he represents leadership of a nationalistic, bourgeois-democratic revolution which precedes a Communist rise to power." His brother, Raúl, & National Bank Pres. Ernesto (Che) Guevara were "strong friends" of the Communist Party.

Cabell estimated that there were 240,000 Latin American Communists (10% more than in 1958) and 650,000-750,000 sympathizers.

---

been employed at the base. Among 3,700 Cubans employed there, only 8 Cuban army men could be found, and none was wanted for counter-revolutionary activities, the note said.

The 3d note said U.S. Agriculture Department inspectors had been withdrawn because the Cuban government had neglected to ask for their continued presence despite a series of reminders.

Bonsal returned to Cuba after the Cuban government had given assurances that no charges were pending against him.

Munitions Ship Blown Up. The French freighter La Coubre, loaded with explosives & ammunition purchased from Belgium, was torn apart by a series of blasts at Havana's Talia Tiedra dock Mar. 4. An estimated 75-100 crewmen, dock workers & passers-by were killed in the blast and 200 persons injured. Warehouses on the dock were burned.

Castro, in a funeral oration for 27 dock workers killed, implied Mar. 5 that the explosion had been caused by U.S.-directed sabotage. He said the U.S. consul general & military attaché in Belgium had called on the Belgian government in efforts to prevent the sale of the ammunition.

"We have the right to believe," Castro said, "that those who did not wish us to receive arms . . . are among those guilty of this sabotage." "We do not have proof," he added. (U.S. State Department spokesman Francis W. Tully said Mar. 5 that the U.S. had "indicated to a number of friendly governments our concern over arms shipments to the Caribbean.")

Cuban naval attaché Filipe Vidal Santiago, who arrived in Bogotá Mar. 8 to seek political asylum, said the ship explosions had been caused by hot harbor water. The water's temperature had been raised by hot water emptied into it by a Havana electric plant near the pier, he said. Cuban dock workers who met in Havana Mar. 8 also expressed doubts that the explosions were sabotage.

Cuban Chargé D'Affaires Enrique Patterson was summoned to the office of State Secy. Herter in Washington Mar. 7 and told that Castro's implication that the U.S. was involved in the explosion was "baseless, erroneous & misleading." Castro's statements could only lead to the further "unhappy deterioration" of U.S.-Cuban relations, Herter said.

Herter's protest was rejected as "aggressive & insulting" in a Cuban note delivered Mar. 8 by Foreign Min. Raúl Roa to U.S. Chargé D'Affaires Daniel Braddock in Havana. The note demanded that the U.S. government, "whenever it addresses itself to the representatives of the Revolutionary Government, do so with absolute respect . . . without descending to offensive utterances of a personal character."

At a Washington news conference Mar. 9, Herter said the U.S. had not contemplated breaking off diplomatic relations with Cuba but that future circumstances might make this necessary. He conceded he had spoken "quite severely" to the Cuban chargé but denied that he had been insulting.

In an impromptu sidewalk interview with foreign newsmen Mar. 12, Castro denied that he had accused the U.S. of direct participation in the explosions. The sabotage had not been directly ordered, he said, "but we have proof that American pilots burned our cane fields, bombed our sugar mills . . . and we have the right to doubt the goodwill of the U.S."

Fire Bombings. Cuban officials reported fire-bomb plane raids Jan. 19, 28, 29 & 31 on cane fields & sugar mills at various places on the northern side of Cuba. Radio commentator José Pardo Llada, a friend of Castro's, said Jan. 29 that the raiders undoubtedly came from U.S. territory. American officials said the plane in a Jan. 28 raid was thought to have borne Moroccan registry markings.

A Piper Comanche plane crashed 100 miles east of Havana Feb. 18 while allegedly attempting to fire-bomb sugar-cane fields. Both occupants were killed. Castro, in a TV address Feb. 19, identified the dead men as Robert Kelly of the U.S. Civil Air Patrol & Robert Ellis Frost of Portland, Ore.

The U.S. State Department acknowledged Feb. 19 that the plane had taken off from Tamiami Airport near Miami. State Department press officer Lincoln White said that U.S. Chargé D'Affaires Braddock had been instructed to express "this government's sincere regrets that the plane managed to escape the vigilance of our intensified airport patrols."

U.S. Pilots Captured. A U.S. 2-engine Piper Comanche 250 was shot down Mar. 21 by Cuban soldiers 70 miles east of Havana on the coastal highway between Veradero Beach & Matanzas. Its pilot, Howard Lewis Rundquist, 35, of Miami, was wounded in the leg by gunfire and taken to an army hospital. Its co-pilot, William J. Shergales, 33, of West Hollywood, Fla., was taken to military intelligence headquarters in Havana.

Soldiers also captured Lt. Col. Damasco Montesino, Havana chief of motor patrol during the Batista regime, his daughter Gladys, her husband, Santiago Rodriguez Rodriguez, and Miguel Sánchez Martinez, all of whom allegedly were waiting to be flown out of Cuba to Florida.

A U.S. State Department spokesman Mar. 29 asserted that Shergales had given U.S. consular officials in Havana an affidavit saying he had been hired by Cuban officials to make the flight in order to embarrass the U.S. The affidavit said Shergales had gone to Havana in February and had conferred with Fidél Castro's personal secretary, Dr. Juan Orta, Capt. Alfred Guerra, a Castro aide, and Maj. Abelardo Colome, chief of army intelligence. Orta denied the charge Mar. 29.

U.S. Pilot Killed. Matthew Edward Duke (Ducko), 45, of Palm Beach, Fla. was ambushed & shot to death by Cuban forces May 12 as he maneuvered a rented U.S. 2-engine Piper Apache plane for a landing on a highway 10 miles west of Havana.

4 men & a woman waiting to be flown out of Cuba were arrested. They were: Ex-Secy.-Gen. Francisco Aguirre Fidaulet of the Hotel & Restaurant Workers Federation, who had been imprisoned for misappropriating union funds; Capt. Jacinto Macias Miranda, ex-aide to Gen. Francisco Tabernilla, who had been army chief under Batista; Angel Rolando Rodriguez, ex-Capt. Julian Guillen Perez & Cecilia Ramos Alcandy.

Cuban authorities said Duke had landed near the same spot and flown 3 Cubans out of the country May 8.

U.S. immigration officials revealed May 12 that Duke had b˹˼n one of 22 persons who had been served Dec. ˳. 1959 with formal notices barring them from flights to Cuba. Duke had told the plane rental company in Baton Rouge May 7 that he was flying to Texas.

Eisenhower Scores Regime. Pres. Eisenhower said in a letter Apr. 8 that friends of Cuba considered actions of its "present leaders" a "betrayal" of ideals they had expressed when they won control of the country. The letter, drafted by the State Department with Mr. Eisenhower's personal participation, was signed by U.S. Amb.-to-Chile Walter Howe and delivered by Howe on the President's behalf in reply to a long list of questions Chilean student groups had given the President in Chile Feb. 24.

The U.S. letter said Cuba's friends "have been gravely disillusioned by what is coming to be considered a betrayal of these ideals in such matters as freedom of expression, equal protection of the laws & the right to choose a representative government." The letter denied that U.S. officials had "made any public statements or committed any acts which may be reasonably construed as unfriendly toward the Cuban government & people." It charged that "the leaders of the Cuban government . . . seem to have intentionally made derogatory & hostile statements" about the U.S. "government & people, designed to disrupt our traditionally amicable relations." The message reiterated that the U.S. did not oppose Cuban agrarian reform but held that its attainment "is not furthered" by Cuba's "failure . . . to recognize the legal rights of U.S. citizens who have made investments in Cuba."

Dorticós' Reply. Pres. Osvaldo Dorticós Torrado denied in a 15-page letter to the Chilean Student Federation Apr. 12 that Cuba had assumed a hostile attitude toward the U.S. The Cuban government, he said, had limited itself "to carrying out a program of achievement designed to win the full development of our nation." He rejected the charge that Premier Castro had betrayed the revolution. He said that "international strategies are set up against Cuba, attempts are made to isolate us internationally and efforts are concerted against the Cuban revolution."

U.S. Charges Slander. In an aide mémoire delivered to the Foreign Ministry in Havana June 4, the U.S. State Department charged the Cuban government with an "official campaign of slander" against the U.S.

The note: (a) denied an assertion by Pres. Dorticós in Montevideo that U.S. citizens in Cuba had been "reimbursed" for expropriated property; (b) said "not a single American property owner has been reimbursed"; (c) noted that while unpaid balances due to U.S. exporters had accumulated to more than $100 million in the past year, enough money had been found to buy armaments worth "tens of millions of dollars" and to increase total dollar reserves; (d) denied as "absurd" a charge by Premier Castro that the U.S. desired to cause serious harm to Cuba because it had decided to end technical aid to Cuba; (e) rejected the "often repeated" theme that close economic cooperation between the U.S. & Cuba had worked to the detriment of Cuba's economy & well-being; (f) charged that certain Cuban officials had not cooperated with the U.S. to prevent unauthorized flights to Cuba and instead had "conspired & connived" to violate U.S. laws "designed to protect Cuba from such incursions"; (g) protested against the distribution of a pamphlet imputing to the U.S. responsibility for the Mar. 4 Havana ship explosion; (h) noted that a May 14 request for an explanation of why a Cuban patrol boat had fired on a U.S. vessel was still unanswered.

The U.S. note was "categorically rejected" by the Cuban Foreign Ministry June 4. Castro, in a speech before the National Congress of Barbers & Beauty Shop Operators June 8, also rejected the U.S. note as "aggressive & written in crude language." He said: "If they [the U.S.] intend to strangle our national economy they will fail. . . . If they try to invade us they will be defeated."

The U.S. State Department protested June 8 to Cuban Delegate-to-OAS Carlos Lechuga against the "continued dissemination of baseless charges," by Castro, seeking to implicate the U.S. in the March explosion of a French munitions ship in Havana.

Castro rejected the U.S. note as "aggressive" in a TV speech June 10. He said the U.S. had refused to give assurances that aggression against Cuba was not planned "because they don't want to put themselves under obligation before world opinion." He charged that Chester Lacayo, a Nicaraguan exile, recently had conferred with State Secy. Herter & Asst. State Secy. Roy R. Rubottom about launching an invasion of Cuba from Nicaragua. Cuba announced June 14 that Lacayo and 13 other persons including U.S. pilot Leslie Bradley & ex-Dominican Gen. Miguel Angel Rodriguez Alcantera had been arrested 3 weeks before & charged with planning the invasion.

Aides Ousted. 2 U.S. Embassy legal attachés were expelled June 16 by the Cuban government. In retaliation, the U.S. State Department expelled 2 Cuban consular officials June 18.

In a note June 16 to U.S. Amb. Philip W. Bonsal, Foreign Min. Raúl Roa accused Edwin L. Sweet of Abilene, Tex. & William G. Friedmann of Stillwater, Okla. of "activities that gravely affect the national sovereignty and involved flagrant intrusion in the internal affairs of Cuba." The pair, both ex-FBI men, had been arrested & held for 2½ hours June 15 by Cuban army intelligence officers for "conspiring with counter - revolutionists." E m b a s s y counselor Daniel M. Braddock had protested the arrests, and Roa's note had been in response to the protest. It was disclosed June 16 that Sweet & Friedmann were Justice Department aides responsible for liaison with Cuban police in cases involving persons wanted for crimes in the U.S.

The U.S. expulsion order against the Cubans charged them June 18 with espionage, racial agitation & "highly improper acts incompatible with their status as consular officials." State Dept. press officer Joseph Reap identified the 2 as Carlos Manuel Lazaro Felix Sanchez y Basquet, 35, assistant consul in Miami, and Dr. Berta Louise Pla y Badia, Cuban cultural attaché in New York. He described Sanchez y Basquet as the "principal Cuban intelligence agent in the Miami area, and accused Dr. Pla y Badia of "distributing anti-U.S. propaganda . . . to augment racial dissension in the U.S." Cuban Chargé d'Affaires Henrique Patterson protested the expulsions as "unjust" and a reprisal against Cuba's earlier action.

U.S. Lands Seized. The Court of Social & Constitutional Guarantees May 12 upheld the right of the Agrarian Reform Institute to seize lands owned by the U.S. government-built Nicaro Nickel Co. (valued at $110 million) & the Moa Bay Mining Co. (Freeport Sulphur Co. subsidiary valued at $75 million) in Oriente Province. Government seizure of Moa Bay's nickel & cobalt plant was announced by Pres. Dorticós Aug. 14. The Moa Bay company had been shut down since Apr. 8, when all but 100 of the plant's 1,000 employes were dismissed. 2,800 employes of the Nicaro plant took over its operations Oct. 2 after the U.S. informed Cuba the plant would be closed because of harassment & "confiscatory" taxes.)

The government announced May 16 that expropriation of the United Fruit

Co.'s 271,410 acres (valued at $32 million exclusive of tools & machinery) had been completed. Capt. Antonio Nuñez Jiminez, head of the Agrarian Reform Institute, said at a mass meeting in Mayari May 16 that the company would be repaid with only $6 million worth of 20-year bonds "because ours is not a government of fools."

Oil Refineries Seized. The Texas Co.'s petroleum refinery in Santiago de Cuba was seized June 29 by the Cuban government after the company had refused a government order to refine Soviet-purchased oil.

U.S. & British-owned refineries in Cuba had rejected the demand June 7 on grounds that they could not process oil from sources other than their own (mostly Venezuelan). At that time Cuban Petroleum Institute officials had declared (reported June 9 by the Wall St. Journal) that the oil companies' position did not "justify retaliation by the government." But Premier Castro, in a June 10 speech, had warned the oil companies to "reconsider" their refusal or "accept the consequences." The Cuban government also had refused to allow the oil refineries to use U.S. dollar exchange to pay for their Venezuelan petroleum.

The Soviet petroleum had been obtained under terms of a commercial agreement, signed in February, providing for Cuban shipment to the Soviet Union of 5 million tons of sugar in the next 5 years.

A U.S. State Department statement June 29 called the Texaco seizure a "naked . . . contravention of the norms of conduct by responsible governments," and warned that the importing of Soviet oil might do "substantial" injury to the economy of Venezuela.

Cuba seized the Esso (Cuba), Inc. (Standard Oil of N.J. subsidiary) & the Shell Petroleum Co. (Royal Dutch Shell Group) refineries July 1. The 3 oil refinery seizures extended direct government control to all oil processing in Cuba. The refineries were taken over by the Cuban Petroleum Institute after the companies had refused to process Soviet crude oil.

In a note delivered to the Ministry of Foreign Relations July 5, the U.S. called the seizures illegal, arbitrary &

unethical and asked that they be reversed. The U.S.-owned refineries were valued at $48,200,000. An additional $50 million was owed to them for foreign exchange they had advanced to pay for Cuban imports of crude oil.

British Amb.-to-Cuba Stanley Fordham July 5 delivered a note protesting the seizure of the Shell refinery. The note said Shell had continued to buy crude oil from Venezuela "for over a year" despite the Cuban government's refusal to allow it to make payments. The note said $17 million was owed to crude oil suppliers as of June.

Sugar Threat. Premier Castro, in a speech June 29 to thousands of workers & peasants, threatened to seize all U.S.-owned sugar mills if the Cuban sugar quota were reduced by the U.S. Congress. Shouting angrily against "Yankee imperialists," Castro declared that all "Yankee property down to nails in their shoes" would be seized, "if they try to force us to surrender by hunger."

U.S. Cuts Sugar Quota. In retaliation for the Castro regime's anti-U.S. policies, the U.S. Congress voted July 3 to give the President discretion to cut the quota for Cuban sugar exports to the U.S. to any level he considered to be in U.S. interests.

This authority, expiring Mar. 31, 1961, was granted in a clause of a bill (HR12311) extending the Sugar Act of 1948 for a year beyond its current Dec. 31, 1960 expiration date.

Pending action by Pres. Eisenhower, the U.S. Agriculture Department July 5 suspended imports of Cuban sugar.

Mr. Eisenhower July 6 ordered a 700,000-ton cut in Cuba's 3,119,655-ton 1960 quota. Since 2,379,903 tons of the quota already had been shipped or certified for shipment, the President's action reduced permissable Cuban sugar exports to the U.S. for the rest of 1960 to 39,752 tons. Mr. Eisenhower simultaneously canceled an additional 156,000-ton share of the 1960 U.S. market that was to have been alloted to Cuba to help make up deficits in quotas of other suppliers. Pres. Eisenhower, assailing the Castro regime's "deliberate policy of hostility" toward the U.S., said he made the cuts with "the most genuine regret." He said Cuban commitments to pay for Soviet goods with Cuban sugar—"traded at prices well below those which it has obtained in the United States"—had raised serious doubts as to whether the U.S. could depend on Cuba as a source of sugar.

Quota-cutting authority had been requested by State Secy. Herter June 22 in an appearance before the U.S. House Agriculture Committee. Before testifying in secret, Herter said publicly that such authority could be used by the U.S. to "reduce the dependence of its consumers on Cuban sugar, the supply of which may become increasingly uncertain." He said experts foresaw a 1,000,000-ton drop in Cuba's annual sugar production.

(Castro declared in a June 24 TV address that Herter had lied in reporting a decline in Cuban sugar output. Castro offered to give up Cuba's full share of the U.S. sugar market if Cuban sugar production dropped 1,000,000 tons in 1961. But he conditioned this offer on U.S. agreement to increase the Cuban quota by 1,000,-000 tons if Cuban sugar output rose by that figure. "We can sell sugar to the United States at a cheaper price if they want it," Castro added.)

3 weeks before Herter testified, the House committee, by a straight party-line vote, had approved a Democratic measure extending the Sugar Act for 4 years and barring the reduction of any country's sugar quota for any reason other than inability to fill it. This Democratic proposal would have meant a 156,000-ton increase in Cuba's 1960 quota and a 165,000-ton rise in Cuba's 1961 quota. The former increase would have been Cuba's share in making up deficits in Puerto Rican, Hawaiian & Virgin Islands quotas. The latter would have been Cuba's share in filling the normal rise in U.S. demand.

The Democratic measure was denounced by Rep. William E. Miller (R., N.Y.) as "pro-Castro" & a "bonus for communism." Miller urged support of a proposal of Rep. Charles E. Hoeven (R., Ia.) to extend the Sugar Act for one year and authorize the President to cut Cuba's quota.

The Hoeven measure originally was opposed by Rep. Harold D. Cooley (D., N.C.), committee chairman, as "overt economic aggression." But Cooley changed his position against quota-cutting authority June 22 after hearing Herter's secret testimony.

The Hoeven proposal was approved by a 394-0 House vote June 30 but ran into trouble in the Senate, which voted by 84-0 July 2 to let the President cut Cuban sugar quotas for only the rest of 1960. The House, in a fight led by Cooley, refused July 2 to accept the Senate restriction.

The final compromise, granting the quota-cutting authority until Mar. 31, 1961, was worked out by Senate-House conferees early July 3 after an all-night session. It was approved by unanimous House voice vote & 32-24 Senate roll-call vote July 3.

(Pres. Eisenhower Dec. 16 extended the ban on purchase of Cuban sugar through Mar. 31, 1961. The ban applied to Cuba's entire 824,299-ton quota during the Jan. 1-Mar. 31 period. Mr. Eisenhower said he had acted because of Cuba's policies of "deliberate hostility" toward the U.S. & of committing "increasing amounts" of sugar to Communist countries.)

Castro, in a June 23 address, denounced the U.S. quota-cutting move as "economic aggression" and warned that U.S. "economic aggression" would be met by the seizure of all U.S. property in Cuba.

U.S. Fills Sugar Supply Gap. The U.S. Agriculture & State Departments, in a move to help fill the 760,-000-ton sugar supply gap created by the virtual elimination of Cuba as a supplier, authorized 6 other nations July 14 to add 137,234 tons to their 1960 quotas.

Seizures Protested. U.S. protests of July 16 & Aug. 8 against seizure & nationalization of U.S. property in Cuba were rejected by Cuba Aug. 13.

The July 16 note said the Nationalization Law of July 6 was "discriminatory," that it applied only to U.S.-owned property and that it was in retaliation for U.S. reduction of the Cuban sugar quota. The note said that should the law be used to seize U.S. property, it would be regarded by the U.S. as "further evidence . . . of a pattern of economic & political aggression" against the U.S. (The U.S. State Department estimated the value of U.S.-seized property in Cuba at between $700 million & $1 billion.)

The nationalization decree applied to all 36 U.S.-owned sugar mills, the Cuban Electric Co. (owned by the American & Foreign Power Co. of New York), the Cuban Telephone Co. (owned by International Telephone & Telegraph Co. of New York) & Standard Oil & Texas Co. refineries.

Soviet-Bloc Sugar Deals. The U.S. Agriculture Department reported July 3 that Soviet-bloc countries had bought no Cuban sugar in Jan.-Apr. 1959 but had bought 763,715 tons in Jan.-Apr. 1960. The USSR imported 602,559 tons, Poland 94,616, Red China 51,861 & East Germany 14,679. By comparison, U.S. imports of Cuban sugar rose from 1,274,000 tons in Jan.-Apr. 1959 to 1,391,000 in Jan.-Apr. 1960. But the USSR had agreed Feb. 13 to buy 1,000,000 tons of sugar annually through 1964.

Cuban Pres. Dorticós announced July 10 that Soviet Premier Krushchev had promised to buy the 739,752 tons of sugar cut from planned U.S. 1960 sugar imports "if the Cuban government finds difficulty in selling it."

In a treaty signed July 23, Communist China agreed to buy 500,000 tons of Cuban sugar annually for 5 years at the world market price of about 3.25¢ a pound (2¢ less than the U.S. purchase price). 20% of the payment was to be made in dollars, the rest in Chinese merchandise. (The treaty was negotiated by Chinese Deputy Foreign Trade Min. Lu Hsu-chang, who arrived in Havana with a 14-man trade mission July 15, and Maj. Ernesto Guevara, president of the National Bank of Cuba. It also provided for the export of Cuban fruits & minerals in exchange for Chinese industrial products, rice, oil & cotton. In addition, Red China extended Cuba credit for the purchase of entire factories.)

Castro said Dec. 19 the USSR had promised to buy 2,700,000 tons of Cuban sugar at 4¢ a pound in 1961. Communist China & other Communist bloc nations, he said, had pledged to buy 1,000,000 tons & 300,000 tons, respectively, at the same price.

(The United Arab Republic promised July 5 to buy 53,000 tons of sugar.

(Under a 3-year Cuban-Japanese trade agreement concluded in Tokyo

Apr. 18, Japan was to buy 450,000 tons of Cuban sugar and export $20 million worth of goods to Cuba annually. The sale of 100,000 tons of sugar to Britain at 2.99¢ a pound & of 50,000 tons to Poland at 2.93¢ a pound, compared with the 3¢-a-pound world market price, had been reported in Havana Mar. 16. These sales increased Cuban 1960 sugar sales to 2,074,000 tons, 97% of its world market quota.)

U.S. Bars Loans for Sugar. U.S. government sources made known in Washington Aug. 19 a new U.S. policy forbidding foreign countries to purchase Cuban sugar with U.S. loans. U.S. officials said the harassment of U.S. ships in Cuban ports made it impossible to fulfill there the foreign aid agreement requirement that U.S. vessels carry 50% of the goods bought in other countries with U.S. aid funds. The new policy already had barred Morocco from using a U.S. loan to buy $10 million worth of Cuban sugar.

USSR Vs. Monroe Doctrine

Khrushchev's Rocket Threat. Soviet Premier Khrushchev warned July 9-11 that the USSR would support the Castro regime with rockets against U.S. intervention in Cuba and that the USSR no longer considered the Monroe Doctrine a valid basis for what he denounced as U.S. domination of Latin America.

In a televised Kremlin speech to a Soviet teachers' convention, Khrushchev asserted July 9 that the USSR would back Castro against "the intrigue of the American imperialists" and that "figuratively speaking, in case of necessity, Soviet artillerymen can support the Cuban people with their rocket fire." Khrushchev said that the U.S.' reduction of Cuban sugar imports was intended "to strangle the economy of Cuba" and could be a prelude to military intervention against Castro. He said the USSR had "rockets which can land precisely on a . . . target 13,-000 km. away," and "this . . . is a warning to those who would like to solve international problems by force & not by reason."

(Tass Oct. 28-29 published 2 versions of a clarification by Khrushchev of his statement promising to aid Cuba with rockets. Khrushchev's clarifica-

tion was made in Moscow Oct. 22 in an interview with Carlos Franqui, director of the Cuban newspaper Revolución. In a summary of the interview, Tass Oct. 28 quoted Khrushchev as saying he would like the promise to defend Cuba to be regarded as a "symbolic" offer. Tass added: "For this purpose Khrushchev said it is essential that the imperialists' threats of [military] intervention against Cuba should not materialize . . . and then there will be no need to confirm the reality of our statement about armed assistance to...Cuba...."
In the full text of the interview, released by Tass Oct. 29, Khrushchev was quoted as agreeing with Franqui that "if this [military] threat is carried out [against Cuba], [Soviet] rockets are adequately prepared" to come to Cuba's defense.)

'Ike' Warns Vs. Soviet Move. In a statement issued July 9, Pres. Eisenhower warned Khrushchev that the U.S. would uphold commitments to prevent foreign nations' "interference" in Latin America and would not "permit the establishment of a regime dominated by international communism in the Western Hemisphere."

The statement, read to newsmen by White House Press Secy. James C. Hagerty, asserted that Khrushchev's rocket threat had been revealing in that (1) "it underscores the close ties that have developed between the Soviet & Cuban governments," and (2) "it . . . shows the clear intention to establish Cuba in a role serving Soviet purposes in this hemisphere." Mr. Eisenhower said: "The Inter-American system has declared itself on more than one occasion . . . opposed to any such interference. We are committed to uphold these agreements. . . . [The U.S.] will not be deterred from this responsibility by the threat Mr. Khrushchev is making."

Khrushchev Vs. Doctrine. Khrushchev said at a Kremlin news conference July 12 that "the Monroe Doctrine has outlived its time" and "has died . . . a natural death." It was the Monroe Doctrine, Khrushchev charged, that had permitted the U.S. to perpetuate the reign of colonialism & monopolies in Latin America.

Khrushchev declared that Mr. Eisenhower's statement linking the Cuban

revolution with communism was "very useful to us" because "we always . . . say that our sympathies are with the colonial peoples, . . . against the exploiters, against the monopolies." Khrushchev said that "should the United States imperialists undertake aggressive action against the Cuban people upholding their national independence, we would support the Cuban people."

Khrushchev made clear that this support would extend to Cuban efforts to evict the U.S. from its naval base at Guantánamo Bay, but he denied as "silly fabrication" the idea that the USSR itself wanted military bases in Cuba. The USSR's "best base is the Soviet Union, from where Soviet rockets can hit unerringly any sector in any part of the globe," he declared.

OAS Gets U.S.-Cuba Dispute

U.S. Memo to OAS. The U.S. State Department June 27 filed an 11-page memo with the Inter-American Peace Committee of the OAS, again accusing Cuba of contributing to Caribbean tensions by a policy of slander against the U.S. The memorandum, made public June 29, charged that the Cuban government had "for many months conducted an intense campaign of distortions, half truths & outright falsehoods" against the U.S. As part of this campaign, the note cited Cuba's (a) charge of U.S. complicity in the French ammunition ship explosion in Havana, (b) attack on a U.S. submarine, (c) encouragement of illegal flights from the U.S., (d) continued anti-American propaganda.

UN Sends Case to OAS. The UN Security Council met in New York July 18 to consider a Cuban complaint of U.S. "economic aggression." The Council heard the Cuban charges & U.S. reply July 18. In a bitter dispute July 19 between U.S. Amb.-to-UN Henry Cabot Lodge & Soviet delegate Arkady A. Sobolev, who supported the Cuban accusations, Sobolev warned that USSR "military might" might be used on Cuba's behalf.

The Council July 19 then adopted an Ecuadorian-Argentine resolution under which it referred the case to the Organization of American States and "adjourn[ed]" its own debate on the issue while it awaited the OAS report. The resolution was approved by a 9-0 vote (the USSR & Poland abstaining).

The resolution was part of a formula proposed July 15 by Dr. José A. Correa of Ecuador, Security Council president for July, & Dr. Raul J. Quijano, deputy Argentine representative. Under the formula, the Cuban complaint was entered on the Council agenda without objection July 18 with the understanding that debate would be limited to the opening Cuban & U.S. statements and that the issue then would be submitted to the OAS. Although the U.S. had the votes to keep the complaint from the agenda and although it had insisted that the dispute really was an OAS rather than a Security Council matter, the U.S. agreed to the formula.

Cuba's request for Security Council action had been made July 11 by Cuban Foreign Min. Raúl Roa, in a letter to Correa. Roa charged that "the repeated threats, harassments, intrigues, reprisals & aggressive acts" to which the U.S. had "subjected" Cuba constituted a "manifest danger to international peace & security."

In his opening statement before the Council July 18, Roa said the U.S. had committed "economic aggression" by reducing Cuba's quota for sugar exports to the U.S. "Better honor without quota than quota without honor," Roa declared. He asked whether it was "necessary, in order to preserve hemispheric unity, democratic principles & continued security, to lower one's head like a submissive bull to the butcher in the abattoir."

Roa charged that the U.S. had sought to distort the character of the Cuban revolution by making it appear, through statements of U.S. officials & private individuals, that Cuba was an "appendage of the Soviet Union." He accused the U.S. of harboring Cuban "war criminals," of allowing U.S. territory to be used as a base for plots & flights against the Cuban government, of exhuming the Monroe Doctrine "with the idea of hegemony" and of taking "such a bellicose tone . . . that it provoked the paralyzing [Soviet] warning."

U.S. Amb.-to-UN Lodge replied July 18 that the U.S. had used no "threats, harassments, intrigues, reprisals or aggressive acts" against Cuba, that the U.S. had no aggressive designs toward Cuba, that the sugar quota cut was necessary to insure U.S. consumers a sure supply, that the U.S. government had made no allegations of Cuban-Soviet ties, that Cuba never had asked the U.S. to extradite accused war criminals, that the U.S. had tried to prevent planes from flying illegally to Cuba and that "we are not frightened by [Soviet] Chrmn. Khrushchev's threat of rockets, and we will live up to those [Western Hemisphere defense] treaties just as we will live up to our [UN] obligations."

Soviet Delegate-to-UN Sobolev, in his July 19 address to the Council, warned U.S. "leading circles" that "the Soviet people will not remain indifferent if an armed intervention is undertaken against Cuba. . . . The Soviet Union, using its military might, will assist Cuba if Cuba asks for assistance." "We are not threatening the United States with our rockets," Sobolev asserted. "We merely say: Do not touch Cuba. Leave Cuba alone. Let it do what it deems proper in its own country, and do not threaten Cuba with your might because other countries also have much might. . . . That is our policy toward the Latin-American countries & toward Cuba."

Lodge replied: "Do not touch us. Do not touch those with whom we are tied. Do not seek to extend Communist imperialism."

Correa, backing Lodge, told Sobolev that the U.S. had made a good record in Latin America, that Latin Americans would "struggle for non-intervention against any attempt to violate it," especially if the attempt were made by "a far distant power." He added that "our [Latin American] efforts for political independence and for economic & social independence . . . are & always will be made by ourselves. . . . We do not accept any kind of guidance. We do not believe in having happiness imposed upon us."

OAS Agrees to Meet. The OAS' 21 representatives voted unanimously in Washington July 18 to hold a foreign ministers' meeting to consider "threats" to continental solidarity, regional defense & democratic principles. The meeting had been proposed by Peru at least a week earlier because of the U.S.-Cuban dispute.

A discussion of Western Hemisphere subversion & Caribbean tensions was approved over Cuban objections as the OAS Council, meeting in Washington, adopted by 20-1 vote Aug 8 a proposed 4-point agenda for the foreign ministers' emergency session.

Agenda items most opposed by the Cuban delegate, Amb. Carlos Lechuga, were: No. 2, calling for "inter-American cooperation . . . for the defense of the democratic American institutions against the subversive activities of any organization, government, or their agents, directed against such institutions," and No. 4, providing for "consideration of existing international tensions in the Caribbean region to assure the harmony, unity & peace of America."

The other 2 items: No. 1, "strengthening of continental solidarity & the Inter-American system especially against threats of extra-continental intervention which may affect them"; No. 3, "consideration of the economic & social factors which cause the political instability in the hemisphere & the intensification of collective action to promote elevation of the standards of living of the underdeveloped regions of the Americas."

Lechuga voted against each agenda item and proposed substitutes accusing the U.S. of "aggression."

Peace Unit Hears Charges. The Cuban government charged in a 24-page statement read before the OAS's Inter-American Peace Committee Aug. 2 that the U.S. had employed "economic aggression" & had fomented "counter-revolutionary" activities in an effort to overthrow the Castro regime.

A U.S. State Department memo to the committee Aug. 7 charged that the Cuban government was "now in open league with the Soviet Union & Communist China" and warned that "the developing military relations between Cuba & the Soviet Union" were the "gravest danger" facing the Americas.

The U.S. statement said that Cuba's regime was a dictatorship, that opposition was suppressed, that the Communist Party (the Partido Socialista Popular) had become the dominant political party in Cuba, that more than 600 persons had been executed & 3,000-6,000 imprisoned for political reasons in Cuba by the end of 1959.

San José OAS Meeting

Caribbean Tensions Noted. The OAS' Inter-American Peace Committee warned Aug. 16, as the OAS foreign ministers opened their emergency meeting in San José, Costa Rica, that "the activities presently contributing to international tensions in the Caribbean . . . present a more difficult problem to cope with than the armed expeditions with which the Inter-American community was faced a year ago." The warning came in a report considered by the ministers after Costa Rican Foreign Min. Alfredo Vargas Fernandes was elected permanent conference chairman.

The report held that: "The general situation in the Caribbean has . . . deteriorated during the past 12 months." While "the actual threat of invasion from one country in the Caribbean area to another has diminished considerably and does not constitute the principle cause of the current tensions," there were other dangerous activities. The Americas were "faced today with a serious crisis which, although not limited to the Caribbean area, is more clearly revealed there than elsewhere." "The peoples of America" frequently faced (a) "reactionary political forces that repudiate democracy and attach themselves to a system of dictatorial government of military type" & (b) "groups . . . under the influence of extra-continental powers."

The committee reported that activities contributing to the Caribbean tensions included: (a) "illegal flights . . . from one country over the territory of another country for the purpose of dropping propaganda, transporting personnel engaged in political activities & causing property damage"; (b) "the alleged abuse of diplomatic status & privileges"; (c) "denial of the rights & privileges of diplomatic personnel" and even "outright assault against them"; (d) "systematic & hostile propaganda carried out by various governments against others, such as the use of "government-controlled radio stations" in campaigns that defamed "chiefs of state" of foreign countries or that "incited to rebellion."

The committee reported that, as a result of collaboration by Honduras, Nicaragua & Costa Rica, "the crossing of frontiers by armed bands" in the area had been "controlled." Nicaraguan exiles, the report said, had sought to overthrow the Nicaraguan government by means of armed incursions from Costa Rica "supported from abroad." "Apparently this support came in part from Cuba," the report declared.

The committee also charged that Caribbean tensions had been "considerably heightened as a result of an increasing desire during recent months on the part of the Soviet Union & other extra-continental powers to intervene more actively in inter-American affairs."

U.S. Accuses Cuba. The U.S. charged Aug. 22 that Cuba was training "Communist agents & guerrillas to spread the Communist revolution throughout Latin America." The accusation, made in a document submitted to the foreign ministers' meeting Aug. 22, said these activities were being carried out by "an international work brigade" supported by the Soviet-sponsored World Federation of Democratic Youth. The document said the brigade was made up of 200 persons from 4 continents who arrived in Havana in August ostensibly for "manual work."

The U.S. document charged that the Cuban government had: (a) attacked the "remaining vestige of individual liberty . . . —religious freedom"; (b) erected a "Cane Curtain" & isolated the Cuban people from contacts with the outside world by cancelling exit permits as of Aug. 22; (c) supported Latin American revolutionary movements; (d) disparaged the OAS in speeches before the Latin-American Youth Congress in Havana Aug. 7 (when Premier Fidél Castro denounced pro-U.S. Latin American nations as "slaves" & "lackeys" of the U.S. and Foreign Min. Raúl Roa described the OAS as a "ministry of

colonies" of the U.S.); (e) engaged in economic duplicity by first charging that the U.S. reduction of the Cuban sugar quota was "economic aggression" & then announcing (through Commerce Min. Raúl Cepero Bonilla Aug. 13) that it would be "more advantageous to Cuba" if the U.S. stopped buying Cuban sugar.

Colombian Plea To Cuba. In a speech before the foreign ministers Aug. 23, Colombian Foreign Min. Julio César Turbay Ayala called on Cuba to reject Soviet support in its dispute with the U.S. & "to accept . . . the discipline of the inter-American system." He challenged Cuban Foreign Min. Roa to deny that Cuba "seeks to instigate the mutiny of the majority of the American countries against" the U.S., that "in any emergency it feels more closely linked to Russia than to the U.S." & that it was fomenting revolution in Latin America through offers of "active & open assistance." Turbay proposed the creation of a 5-nation committee to help settle the U.S.-Cuban dispute.

Roa Reaffirms Red Ties. Roa told the OAS foreign ministers Aug. 25 that Cuba was determined to maintain its close relations with the Soviet Union. He denied that the promise of Russian military assistance represented Communist intervention in the Western Hemisphere. The threat to American nations, he said, was not a "hypothetical extra-continental intervention, but the actual & present threat of aggression by the [U.S.] government." Roa said Cuba was ready to negotiate its differences with Washington "in a bilateral way on a footing of equality."

Herter replied that Roa's speech showed that (a) Cuba "is walking hand-in-hand with the Sino-Soviet bloc"; (b) ". . . Cuba is willing to assist the Soviet Union in spreading communism in the rest of the hemisphere . . ."; (c) Cuba "exhibited a calculated antagonism [to the U.S.] which will render more difficult the efforts of my country to restore friendly relations with Cuba."

Red Intervention Condemned. A "Declaration of San José," condemning Communist intervention in the affairs of American nations, was adopted Aug. 28 at the final working session of the San José meeting. Latin American foreign ministers, in approving the declaration, supported the U.S. in its dispute with Cuba.

The declaration did not mention Cuba. It was thoroughly understood, however, that the declaration was intended as a criticism of Cuban Premier Fidél Castro for accepting a Soviet offer of military protection against the U.S.

The Cuban delegation walked out of the conference Aug. 28 after the ministers had voted to adopt the first 2 of the declaration's 8 sections. Roa had asserted before the voting began that Cuba "is under a real threat," that Cuba "is absolutely convinced

---

THE DECLARATION OF SAN JOSE
(Partial Text)

"The 7th Meeting of Consultation of Ministers of Foreign Affairs
"1. Condemns emphatically intervention or the threat of intervention . . . from an extracontinental power in the affairs of the American republics and declares that the acceptance of [such] a threat . . . by any American state jeopardizes American solidarity & security. . . .
"2. Rejects, also, the attempt of the Sino-Soviet powers to make use of the political, economic or social situation of any American state, inasmuch as that attempt is capable of destroying hemispheric unity and jeopardizing the peace & the security of the hemisphere.
"3. Reaffirms the principles of nonintervention by any American state in the . . . affairs of the other American states, and it reiterates that each state has the right to develop its culture, political, & economic life freely & naturally, respecting the rights of the individual & the principles of universal morality, and . . . no American state may intervene for the purpose of imposing upon another American state its ideology or political, economic, or social principles.
"4. Reaffirms that the Inter-American system is incompatible with any form of totalitarianism and that democracy will achieve the full scope of its objectives in the hemisphere only when all the American republics conduct themselves in accordance with the principles stated in the Declaration of Santiago, Chile. . . .
"5. Proclaims that all [OAS] member states . . . are under obligation to submit to the discipline of the Inter-American system . . . and that the soundest guarantee of their sovereignty & their political independence stems from compliance with the provisions of the [OAS] charter. . . .
"6. Declares that all controversies between member states should be resolved by the measures for peaceful solution that are contemplated in the Inter-American system.
"7. Reaffirms its faith in the regional system and its confidence in the Organization of American states . . . since it is in this organization that the members find the best guarantee for their evolution & development.
"8. Resolves that this declaration shall be known as the 'Declaration of San José, Costa Rica.' "

---

that she feared an imminent United States military attack" and that Cuba, therefore, had to accept aid from any nation in order to survive.

"We have decided to withdraw from the conference," Roa said, "because Cuba could not find protection in the Organization of American States."

All 19 remaining countries represented at the meeting voted for the declaration although Venezuela & Mexico noted reservations. The conference rejected 4 softening amendments proposed by Venezuela. Venezuelan Foreign Min. Ignacio Luis Arcaya withdrew as head of the Venezuelan delegation because, he said, his personal sympathy for the Cuban revolution did not permit him to sign the declaration. Arcaya was replaced by Marcos Falcon-Briceno, who signed.

Before leaving San José Aug. 29, U.S. State Secy. Herter said the declaration was a "clear indictment of . . . Cuba and particularly the role which it has played in furthering the Sino-Soviet efforts of intervention into this hemisphere."

The Mexican delegation said the declaration was of "a general character and not "a condemnation or a threat against Cuba. . . ."

The conference Aug. 28 voted to established a 6-nation committee to investigate Caribbean disputes, "especially" U.S.-Cuban controversies, and to offer its good offices in settling them. The committee's members: Mexico, Costa Rica, Colombia, Venezuela, Brazil & Chile.

A Cuban proposal to "warn" the U.S. not to continue its "provocation, intervention & aggression" was rejected by an 18-1 vote Aug. 28.

San José Aftermath

Castro Scores U.S. Role. Castro, in 2 Havana speeches Aug. 30, attacked the U.S. role at the OAS conference and said Cuba would continue to depend on Soviet "rocket support." He called the Declaration of San José "an imperialistic United States maneuver" to get Cuba to renounce such aid.

Castro charged that Latin American nations that voted for the declaration had been "bought" by U.S. promises of a share of a proposed $600 million Latin American loan fund.

In a Havana speech Aug. 24 Castro had said Cuba would "triumph with the OAS or without the OAS." He said: "We [Cuba] are friends" of the USSR & Communist China and would remain so "because they have shown that they are our friends, while you [the U.S.] have attacked and wished to destroy us."

UN 'Summit' Session

World Leaders in N.Y. Views on the Cuban situation were exchanged by world leaders in New York at the UN General Assembly's 1960 session, which convened Sept. 20.

The session was turned into an unprecedented world summit meeting by the presence of Soviet Premier Khrushchev and a host of leaders from Communist, neutralist and some Western countries. This transformation was an outgrowth of the failure of a true East-West summit conference in Paris earlier in 1960 and of the breakup of the 10-nation Geneva disarmament conference. Discussions of the Cuban problem formed only a minor part of the proceedings, and nothing of importance concerning Cuba was done.

Khrushchev, Castro and other Communist leaders who attended the UN session were restricted by the U.S. State Department to Manhattan Island (N.Y. City), site of UN headquarters.

Castro arrived in New York by air Sept. 18 at the head of a UN delegation including Cuban Foreign Min. Raúl Roa & Capt. Antonio Nuñez Jimenez, director of Cuba's National Agrarian Reform Institute. Castro was cheered by several thousand Cubans but was hissed by American onlookers as he was driven to the Shelburne Hotel in Manhattan. The Shelburne had accepted the Cubans at UN & State Department request after several Manhattan hotels had refused them accommodations. Castro, however, led the Cubans in a walkout from the Shelburne Sept. 19. He charged that the hotel had made "unacceptable cash demands." He protested to Secy. Gen. Hammarskjöld but refused a UN offer of free accommodations elsewhere and went to the Theresa Hotel in New York's predominately Negro Harlem district.‡

Castro addressed the Assembly Sept. 26 and used his speech to attack the U.S. He said:

Soviet Premier Khrushchev & Cuban Premier Castro embrace at opening session of UN General Assembly Sept. 20 as Cuban Foreign Min. Roa (right, wearing sun glasses) watches. Khrushchev & Castro had met for the first time earlier Sept. 20 when Khrushchev visited Castro in Harlem.   (Wide World photo)

The U.S. hoped to destroy the Cuban revolution on the pretext that it was Communist-dominated; actually Cuba was being punished with "economic aggressions" for its land reform & nationalization measures against U.S. monopolies that had made Cuba a satellite of the U.S.

The U.S. had permitted air attacks on Cuba from its territory and had allowed the use of the Swan Islands, "which belong to Honduras," by "war criminals" for broadcasts & training of an invasion force against Cuba; Cuba would force by diplomatic means a U.S. withdrawal from the Guantánamo Bay naval base if the base became "a threat" to Cuba.

The U.S., which had aided the Batista dictatorship, denied all aid to the revolutionary regime and gave refuge to men who had "murdered hundreds of defenseless peasants"; its $500 million Latin aid program was intended to prevent the spread of the Cuban revolution.

Cuba supported most major Soviet foreign policies, particularly its pro-

‡Castro told the UN Assembly Sept. 26 that he had been subjected to "humiliating treatment" on his arrival in New York. He complained of illtreatment by public authorities and charged the Shelburne with making "extortion efforts" for its services. He said that it was "a humble hotel, a hotel of the Negroes in Harlem, that sheltered us" despite others' hostility.

posals for total disarmament. Cuba shared Soviet opposition to formation of an international peace force that could be used to halt revolutions in the smaller states. It "condemned" Hammarskjöld's policies in the Congo and supported Algeria's war for independence & Communist China's right to membership in the UN.

U.S.-Cuban Charges. A. U.S. document charging the Castro regime with deliberately undermining Cuban relations with the U.S. was submitted to the Assembly Oct. 14 by U.S. Amb.-to-UN Wadsworth.

The U.S.' "fact sheet" contained a point-by-point refutation of what it termed Castro's "untrue & distorted allegations" against the U.S. in his Sept. 26 speech to the Assembly.

Major points in the document:

● The U.S.' acquisition of the Guantanamo naval base was not "imposed by force" but was the result of an agreement negotiated between "sovereign governments" in 1902-03 and freely reaffirmed in 1934.

● The charge that U.S. "monopolists" imposed the regime of Fulgencio Batista & other dictatorships on Cuba was "ridiculous." U.S. foreign policy was based on non-intervention.

● It was Cuba, not the U.S., that first effected sharp reductions in U.S.-Cuban trade. Months before the U.S. reduced Cuba's sugar quota, Cuba had agreed to sell a major part of its crop to Russia & Red China.

● U.S.-owned property in Cuba was not expropriated but confiscated without compensation. This was contrary to international law & Cuban laws.

● The Castro régime had interfered with freedom of religion & forced into exile Cuban newspaper editors opposed to government policies or communism; it had suppressed all political parties except the Communists, denied free elections & arrested thousands for political offenses.

● The U.S. had not "taken over" the Swan Islands claimed by Honduras. They had belonged to the U.S. for almost 100 years, but the U.S. had offered to discuss the claims.

● The Castro regime welcomed the growing intervention in its affairs by the Soviet Union & Communist China.

The Assembly voted by 45-29 (18 abstentions) Nov. 1 to reject a Soviet-supported Cuban demand for immediate debate of alleged U.S. aggression against Cuba. The Assembly voted by 53-11 (27 abstentions) to refer the Cuban charges to the Political Committee for preliminary debate.

The Assembly decision came after a 2-day debate in which Cuban Foreign Min. Roa & Soviet Rep.-to-UN Zorin

Pres. Eisenhower addresses UN 'summit' session in New York Sept. 22.
(Wide World photo)

reiterated charges that the U.S. had committed economic & aerial aggression against Cuba and was conspiring to overthrow the Castro regime. Roa Oct. 31 accused the 2 U.S. Presidential candidates, Vice Pres. Nixon & Sen. John F. Kennedy, of arguing over Cuba as if it were "real estate belonging to the United States."

Cuban-Soviet Ties

Mikoyan Visit. In a commercial-aid pact signed by Castro & Soviet First Deputy Premier Anastas I. Mikoyan in Havana Feb. 13, the USSR agreed to buy 1,000,000 tons of Cuban sugar annually in 1960-64 at world market prices (currently 3¢ a lb.), to give Cuba a $100 million credit repayable in 12 years at 2½% interest and to furnish technical assistance to Cuba during 1961-64. The credit was to be used to buy equipment, machinery & materials. 200,000 tons a year of the sugar was to be paid for in U.S. dollars, and credit for the balance was to be used against Cuban purchases in the Soviet Union.

Replying to a U.S. Embassy statement calling attention to the fact that the U.S. bought 3 million tons of Cuban sugar annually at an average price of 2.77¢ above the world market price, Cuban Commerce Min. Raúl Cepero Bonilla said Feb. 15 that the sugar bought by Russia would be consumed there and not resold on the world market. (The AP reported from Washington Nov. 25 that the USSR had started dumping Cuban sugar at less than world market prices [$98 a metric ton] with the sale to Jordan of 1,500 tons of Cuban sugar for $84 a metric ton.) Cepero Bonilla said almost all Cuban imports came from the U.S. and that the balance of payments always had been unfavorable to Cuba. The price the U.S. paid for sugar, he declared, was "not a favor" but a device to protect its inefficient & expensive domestic cane & beet sugar production.

Mikoyan had arrived in Cuba Feb. 4 to open a Soviet trade exhibition. (Anti-Communists demonstrated against Mikoyan's visit in Havana's Central Park, outside the Soviet exhibition, shortly before he went there with Castro for opening ceremonies Feb. 5. Police fired into the air to

Soviet First Deputy Premier Mikoyan & Castro sign agreement in Havana Feb. 13 for Soviet purchase of Cuban sugar and extension of $100 million credit to Cuba. (Wide World photo)

quell the demonstration and arrested 17 university students who took part in it.) Mikoyan left Havana Feb. 13 for a 2-day visit to Norway. It was reported Feb. 12 that Mikoyan had offered to sell Soviet military aircraft to the Cuban government. When asked Feb. 13 if Cuba would be allowed to buy the planes with the new credit, Mikoyan replied that "the matter [had] not been discussed."

(The appointment of Sergei M. Kudryavtsev as Soviet ambassador to Cuba was reported July 7 by Tass. Kudryavtsev reportedly had been linked with the Soviet spy ring uncovered in Canada in 1946.)

Polish Pacts. Under a Polish credit-barter pact, announced in Havana Mar. 31, Cuba was to receive seagoing ships, fishing vessels, planes, helicopters, steel foundries, lamination mills, sugar processing machinery, chemical & power plants, food processing plants*, machine shops, textile machinery*, road building equipment, tractors† & diesel & electric motors. Poland was to receive sugar, coffee, canned fruits, hides, tobacco, fibers, rayon cord, manganese, nickel & iron ores.

(Cuban Deputy Foreign Affairs Min. Hector Rodriguez Llompart & Polish Deputy Foreign Trade Min. Josef Kutyn signed in Warsaw Dec. 22 an agreement to expand Cuban-Polish trade to $44 million in 1961.)

* It was reported Apr. 2 that Poland had been trying for 3 years to get U.S. credits with which to buy textile & food-processing machinery for Poland's own use.
† Trybuna Ludu, the Polish CP organ, had reported Feb. 8 that lack of parts had immobilized many of Poland's own scarce tractors.

Communist Factories Bought. Capt. Antonio Nuñez Jimenez, director of the Agrarian Reform Institute, revealed July 23 that Cuba had used $84 million of the $100 million Soviet credit granted under the Russian trade pact signed in February to buy 32 factories. 7 were bought from the USSR, 12 from Poland, 8 from Czechoslovakia & 5 from East Germany. 5 of the Czech plants, for the production of hardware & tools, were to be shipped immediately.

North Korean Pact. A cultural exchange agreement with North Korea was signed in Havana Aug. 29 by Actg. Foreign Min. Raúl Cepero Bonilla & North Korean Min. Han Sang Doo. The 2 countries also agreed to exchange ambassadors.

Red China Recognized. The Cuban government formally recognized Communist China & North Korea Sept. 24. The action, taken at a cabinet meeting, ratified a Sept. 9 vote of the Cuban Parliament "to recognize the . . . Republic of China as the only representative of the Chinese people." Premier Castro, attending the UN General Assembly meeting in New York, participated in the cabinet action by phone. Cuba became the first Latin American nation to recognize Red China.

Cuba's intention of recognizing Communist China and severing relations with Nationalist China had been announced by Castro at a Havana rally Castro had called to protest the Cuba-opposed Declaration of San José. During the same speech Castro physically tore up a 1947 military treaty with the U.S. The pact, known as the Treaty of Rio de Janeiro, had been adopted by the Inter-American Defense Conference in 1947. It provided that no U.S.-furnished arms should be used against any signatory nation.

Chinese Nationalist Amb. Liu Yu-wan delivered to the Cuban Foreign Affairs Ministry in Havana Sept. 3 a note breaking relations with Cuba.

U.S. Charges Red Arms. The U.S. State Department charged in a note to the Organization of American States (OAS) Oct. 28 that Soviet-bloc nations had shipped "thousands of tons" of weapons to Cuba "to give armed support to the spread of its [Cuban] revolution to other parts of the Americas." The note, submitted to OAS Secy. General José A. Mora by John C. Dreier, U.S. representative to the OAS, asked the OAS's 6-nation ad hoc committee to investigate the charges. (The committee had been formed by the OAS foreign ministers' conference in San José in August to investigate U.S.-Cuban controversies.)

The U.S. State Department charged Nov. 18 that Soviet-bloc arms shipments to Cuba since Jan. 1, 1959 had totaled "at least 28,000 tons." It said Castro's regime had increased its "military force [to] 10 times the size of . . . ex-Pres. Fulgencio Batista['s]" and was "far larger than any [other] army in Latin America." It charged that Cuba's regular army totaled 40,000 men & its militia more than 200,000. It said Communist arms imports, mainly from the USSR & Czechoslovakia, had increased sharply since the August OAS meeting in Costa Rica had condemned Cuba & extracontinental intervention in the western hemisphere. Red arms imported by Cuba included 8 MiG jet fighters, 10 Soviet helicopters, 40 tanks, 30 rocket launchers, 25 field guns, 55 howitzers, 45,000 Czech automatic rifles, 60 anti-tank guns & 80 anti-aircraft guns, the statement said. Since July, the statement claimed, "at least 12 Soviet ships had delivered arms & ammunition to Cuba." The statement said thousands of automatic rifles, submachine guns & a great quantity of ammunition also had been purchased from non-Communist countries.

Guevara's Tour. Dr. Ernesto (Ché) Guevara, Cuban National Bank president, toured European & Asian Communist countries for 3 months at the head of a Cuban economic mission. The Cuban mission arrived in Prague Oct. 23 and in Moscow Oct. 30. Guevara said in Moscow Nov. 1 that he was seeking more Communist trade & aid to counter what he called the U.S. economic blockade against Cuba.

An economic cooperation agreement was signed in Peiping Nov. 30 by Guevara & Communist Chinese Finance Min. Li Hsien-nien. Under the pact, Red China gave Cuba an interest-free 240 million-ruble ($60 million) 5-year loan, and the Chinese Communists promised to import one million tons of Cuban sugar in 1961. Guevara returned to Moscow Dec. 8 after visiting

Mongolia, North Viet Nam & North Korea.

Guevara said at a Moscow "friend-ship meeting" Dec. 10 that Cuba 'wholeheartedly support[ed]" the manifesto adopted in Moscow by the leaders of 81 Communist parties. Guevara Dec. 19 signed a declaration pledging Cuban support of Soviet Premier Khrushchev's foreign policy, including backing of "national libera-tion movements" in Latin America.

A 5-year trade pact and an agree-ment to exchange economic & techni-cal advisers with East Germany was signed by Guevara in East Berlin Dec. 17. East Germany agreed to grant Cuba long-term credits & to exchange industrial goods for Cuban sugar.

An agreement establishing diplo-matic relations between Hungary & Cuba was signed by Guevara in Buda-pest Dec. 18.

Guevara & Soviet First Deputy Pre-mier Mikoyan signed in Moscow Dec. 19 an agreement providing for the ex-pansion of Cuban-Soviet trade to $168 million in 1961 and for Soviet aid in the building of Cuban industrial proj-ects. The USSR agreed to: (a) supply Cuba's refineries with oil; (b) buy more than the prescribed amount of Cuban sugar in 1961 if the U.S. re-fused to buy any Cuban sugar; (c) build for Cuba an iron & steel mill, an oil refinery, electric power stations & other "industrial enterprises."

Guevara returned to Havana Dec. 23 and said Dec. 26 that under the Dec. 19 trade pact, the USSR would build 100 industrial plants, including cotton mills and textile and machine-tool plants, in Cuba. He said that Commu-nist China would build factories in Cuba and that Soviet-bloc nations would send technicians to Cuba to ex-plore for oil & iron ore deposits & to increase Cuba's steel output from 40,-000 tons a year to 200,000.

U.S. Vs. Czech Air Route. The N.Y. Times reported Dec. 17 that the U.S. had urged the West German, Belgian & Portuguese governments to deny Czechoslovakia transit & landing rights for its prospective Prague-to-Havana air service. Portugal report-edly had refused to let the Czechs refuel & service Cuba-bound planes in the Azores.

Red Becomes Envoy. Dr. Salvador Garcia Aguero, ex-senator & a Com-munist Party leader, was appointed Cuban ambassador to Guinea Dec. 23. He was the first CP official appointed a Cuban ambassador.

U.S.-Cuban Rift Widens

Americans Advised to Leave. The U.S. State Dept. said Sept. 29 that it had urged U.S. Embassy employes in Havana and about 4,000 other U.S. civilians in Cuba to send their de-pendents home because of harassment of Americans by Cuban police. About 500 American wives & their children already had been sent back to the U.S. Embassy aides had started sending their families home in August. A de-partment spokesman said the advice was a precautionary move & was not in anticipation of a break in U.S.-Cuban relations. He said 43 Ameri-cans had been "harassed" by Cuban police since Jan. 1.

The department Sept. 30 urged U.S. tourists & others planning to visit Cuba not to go "unless there are com-pelling reasons."

Bonsal Recalled. The U.S. State Department announced Oct. 20 that Amb. Philip W. Bonsal, 57, had been recalled to Washington for "an ex-tended period of consultations" and that "no date has been fixed" for his return to Cuba. The U.S. Embassy in Havana was to be headed by Daniel M. Braddock, chargé d'affaires.

U.S. Imposes Trade Embargo. The State Department Oct. 20 announced an immediate embargo on most U.S. exports to Cuba to counter the "dis-criminatory, aggressive and injurious economic policies" of the Cuban gov-ernment.

The ban covered technical informa-tion and all goods except medicines. medical supplies & certain foods. It applied to corn products, wheat, rice & soybeans that either were subsi-dized for export by the CCC or that came from CCC stocks. The sale, trans-fer or charter of U.S.-owned ships to the Cuban government or to its nationals was forbidden without prior Maritime Commission approval.

The U.S. Commerce Department said the embargo had been imposed in compliance with the U.S. Export Control Act, which permitted the "use

[of] export controls . . . to further [U.S.] foreign policy."

The State Department c a l l e d Cuba's "discriminatory" economic practices a "deliberate political policy to divert trade from the United States." Practices cited by the department included: "taxes & other restrictions" against U.S. flour, potatoes, rice, drugs, cigarettes, shoes, automobiles & other products; 30%-100% Cuban surcharges on remittances on foreign exchange to pay for certain U.S. imports; "pressures" on U.S. customers in Cuba to divert their orders to other countries; the seizure of U.S.-owned property in Cuba.

(U.S. exports to Cuba had dropped from $546,200,000 in 1958 to $436,706,000 in 1959, the year Castro took over. U.S. exports in 1960, prior to the embargo, were estimated at no more than $250 million.)

A government radio broadcast charged Oct. 20 that the purpose of the embargo was to "soften up" Cuba as a prelude to a U.S. military invasion of the island from Guatemala.

U.S. Sees 'Invasion Hoax.' Cuban Actg. Foreign Min. Carlos Olivares Sánchez rejected Oct. 22 an Oct. 21 U.S. note protesting the reported painting of U.S. flags on "5 or 6 Cuban [transport] planes" at the Cuban Air Force base of San Antonio de los Banos in Havana. A U.S. State Department official said the alleged flag painting might have been part of an "invasion hoax" to fake a U.S. attack against Cuba.

Invasion Mobilization. Thousands of Cuban militiamen Oct. 30 were called out to take up positions in various sections of Havana & its suburbs to prepare for "imminent" U.S. invasion. Hospitals were ordered to reserve beds & set aside emergency supplies to care for wounded.

Foreign Min. Raúl Roa asked the UN Security Council Dec. 31 to meet and investigate his charges that the U.S. was prepared to "carry out, within a few hours, direct military aggression" against Cuba.

U.S. Marines at Guantanamo. 1,450 U.S. Marines landed at the U.S. naval base at Guantanamo Oct. 29 for weekend shore leave during training exercises in the Caribbean. The Navy announced the shore leave Oct. 28 to discount suggestions that the Marine visit might be part of a U.S. "invasion" of Cuba. The White House said Oct. 29 that the Marine visit had Pres. Eisenhower's "full approval." Cuban newspapers Oct. 29 called the Marine visit part of a U.S. plan to conduct a fake attack against Guantánamo as a pretext to invade Cuba.

'Ike' Vows Base's Defense. Pres. Eisenhower declared Nov. 1 that the U.S. would take "whatever steps" necessary to maintain the U.S. naval base at Guantánamo. Mr. Eisenhower's statement, released by the White House, cited the U.S.-Cuban treaty of 1903 that gave the U.S. "complete jurisdiction and control over the area." He said that the treaties governing the base could be changed only by agreement between the U.S. & Cuba and that the U.S. had "no intention of agreeing to modification or abrogation of these agreements. . . ."

The U.S. Navy said Nov. 2 it had placed clearly-marked land mines around its Guantánamo base to ward off a possible attack.

Cuban Pres. Dorticós said Nov. 2 that Cuba "would never commit the stupidity" of attacking the U.S. naval base at Guantánamo. In one of a series of "revolutionary talks" aimed at showing that the U.S. was Cuba's "greatest enemy," Dorticós said Cuba would claim the base "in the proper time & through the proper political procedure." He said Pres. Eisenhower's Nov. 1 statement that the U.S. would defend Guantánamo was a U.S. attempt "to justify an armed aggression" against Cuba.

U.S. Firms Seized. The Cuban government Oct. 25 nationalized 166 U.S.-owned enterprises valued at about $250 million. A decree authorizing the seizures was issued in retaliation for the U.S. embargo on exports to Cuba. It practically erased all major U.S. investments in Cuba.

The U.S.-owned enterprises nationalized included the branches of F. W. Woolworth, Sears Roebuck & Co., General Electric, Westinghouse, International Harvester, Remington Rand, Squibb & Co., Continental Can Corp. and Coca-Cola and several luxury hotels & casinos.

The U.S. Commerce Dept. reported Nov. 13 that Cuba had seized nearly $1 billion worth of U.S. investments so far. This total, the department said, did not include such U.S.-owned assets as the U.S. government's Nicaro nickel plant or investments of

Americans who had become residents of Cuba. The department report estimated that as of 1957 U.S. business in Cuba had: (a) paid about $170 million annually in wages & salaries; (b) paid the Cuban government $70 million in taxes annually; (c) provided jobs for about 160,000; (d) bought nearly $300 million worth of goods & services in Cuba yearly.

(The Cuban government had seized & nationalized branches of these 3 U.S. banks in Havana Sept. 17: First National City Bank of New York, First National Bank of Boston and Chase Manhattan Bank. The Foreign Ministry Sept. 29 refused to accept a U.S. note protesting the seizures. Cuba's nationalization of the Royal Bank of Canada's 23 branches in Cuba [estimated value: $125 million], under "an agreement which the bank accepted," was disclosed by bank officials Dec. 8. Seizure of the food-processing plants of Nestlé Alimentana, the first Swiss company nationalized by Cuba, was disclosed by Switzerland's foreign office Nov. 3.)

U.S. Rocket Pieces Hit Cuba. Pieces of a U.S. satellite-launching rocket fell Nov. 30 over a 20-mile-long area on the Oriente Province coast. The fragments came from a malfunctioning Thor-Able-Star rocket that was destroyed in the air shortly after it was launched from Cape Canaveral, Fla. The U.S. asked the Cuban government Dec. 2 to return the pieces for examination.

U.S. Acts Vs. Export of Revolt

Navy Patrols Trouble Spots. Pres. Eisenhower Nov. 16 ordered U.S. Navy warships & planes to patrol the Caribbean waters off Guatemala & Nicaragua and "seek out & prevent" possible Communist-led invasions. The White House disclosed Nov. 17 that State Secy. Christian A. Herter had received "requests in writing" from Guatemala & Nicaragua for U.S. protection. Both countries had said their governments were jeopardized by Communist-planned invasions from Cuba.

The U.S. State Department said Nov. 17 that the U.S. had acted because of widespread fears that Cuban Premier Fidél Castro might try to extend his revolution to other Latin American countries with the arms supplied him by the Communist bloc. The department said the naval units would provide direct aid in repelling an invasion only if Guatemala & Nicaragua requested and would do so only "within the national jurisdiction of the requesting governments." It was explained that the U.S. naval units would not act against suspected invasion vessels until they had come within 3 miles of the coasts of Guatemala or Nicaragua. The U.S. ships involved were the aircraft carrier Shangri-La and the destroyers Kenneth D. Bailey, Fiske, Hawkins, W. R. Rush & M. C. Fox.

In a Nov. 18 note to UN General Assembly Pres. Frederick H. Boland, Cuban Foreign Min. Raúl Roa charged that the U.S. move was "a flagrant violation of the principle of non-intervention and an act of aggression." He said the U.S. was "arrogating to itself" the role of "an international police force" and had turned the Caribbean Sea "into a private lake."

Foreign Min. Jesus Unda Murillo disclosed Nov. 14 that Guatemala had asked the U.S. for ships & planes to guard its shores against a possible Cuban invasion.

Guatemalan Pres. Miguel Ydigoras Fuentes said Nov. 18 that he had requested U.S. naval aid to provide Guatemala with advance information of an impending attack rather than protection against an attack itself. He said that Mr. Eisenhower's decision was right and that "it came in time and it might dissuade Castro."

Ships & Planes Withdrawn. The White House announced Dec. 7 that Pres. Eisenhower had recalled U.S. Navy warships & planes that had been sent to patrol the Caribbean waters off Guatemala & Nicaragua to block possible Communist-led invasions. The announcement said Guatemala & Nicaragua had informed the U.S. that "the emergency which led to their requests [for the U.S. patrol] had passed."

Peru Charges Plot. Peru broke diplomatic relations with Cuba Dec. 30. The Lima newspaper La Prensa had charged that Cuba had been sending Soviet funds to Peru's Communist Party to finance a revolt. La Prensa published a photo-copy of a letter, allegedly sent to Cuban Amb.-to-Peru Luis Alonso, in which the Peruvian CP acknowledged receipt of such funds from the Cuban Embassy in Lima. La Prensa said the letter was among documents seized by an anti-Castro group in a raid on the Embassy Nov. 8.

Anti-Castro Activity

Rebels Fight in Mountains. 5,000 troops were reported Apr. 15 to have been sent into the Sierra Maestra mountains to combat an insurgent force of about 200 men led by Capt. Manolo Beatón & his brother Cipriano, both ex-members of Premier Fidél Castro's revolutionary movement. Beatón had escaped from a military prison after being arrested on charges of killing Maj. Cristino Naranjo, aide to the late Maj. Camilo Cienfuegos.

Troops converging from Santiago de Cuba, El Caney & Mina del Frio on Beatón's Pico Turquino stronghold were reportedly led by Maj. Raúl Castro, armed forces minister, & by Maj. Juan Almeida, army chief of staff. The anti-rebel operation was directed by Fidél Castro from Las Mercedes, south of Bayamo. Workers' militia units were reported Apr. 12 to have gone into action with Castro's troops to hunt the rebels.

The Beatón brothers & Felipe Martinez were executed by a firing squad June 15 in Santiago de Cuba for high treason, murder & armed revolt. They were the first to be executed under a July 1959 decree of death for counter-revolutionary activities. Beatón had surrendered to 5 peasant militiamen June 10 in Bayamita in the Sierra Maestra. 15 other members of his guerrilla band had been captured previously. A 5-judge summary court-martial had convicted the 3 guerrilla leaders June 14 and had sentenced other members of the band to prison terms ranging from 5 to 30 years. 7 youths were sent to reformatories and one was acquitted.

A rebel group of about 50 men led by Capt. Niño Diaz, a former leading Castro field commander, reportedly operated in Oriente Province also.

In Santiago, once a stronghold of Castro support, an organization called the Movement of Revolutionary Recovery was reported to have gained wide support. Composed almost entirely of Castro's ex-military & civilian companions, the movement was opposed to allegedly Communist & dictatorial influences in Castro's government.

Francisco Gutierrez, a member of the anti-Castro Cuban Liberation Alliance formed in Miami Beach Aug. 7, said at a news conference in New York Sept. 14 that guerrilla forces in the Escambray Mountains numbered 800-1,000. He said they were led by Victor Manuel Paneque, an ex-Castro officer who used the name Maj. Diego. The news conference was attended by Diego and 2 other ex-Castro officers: Maj. Pedro Luís Díaz Lanz, ex-air force chief, & Niño Diaz.

A militia victory over guerrillas in the Escambray Mountains was announced by the government Sept. 26. One rebel had been killed, 3 wounded & 26 captured in a series of clashes, according to the official newspaper, Revolución. UPI quoted rebel sources as saying 19 militiamen had been killed in an ambush.

A 29-day military campaign against the Escambray insurgents resulted in the death of 3 rebels & the capture of 102, the government reported Oct. 9. About 150 others were said to be continuing the fight. 3 members of the government forces were reported killed. The Armed Forces Ministry said that captured arms & ammunition had been parachuted to the rebels Sept. 29 from a "Yankee plane."

At a mass trial of 200 Escambray rebels, 5 leaders were sentenced to die and 20- to 30-year prison terms were imposed on other defendants in Santa Clara Oct. 11. The 5 convicted rebel leaders, who were executed Oct. 12: Plinio Prieto, Senesio Walsh Rio, José A. Palomino, Angel Rodriguez de Sol & Porfirio Ramirez.

Plot Foiled. Troops led by Fidél Castro & his brother Raúl Aug. 16 reportedly smashed a "counter-revolutionary plot" at the North Naval District base in Casa Blanca. The Castro-led soldiers were said to have raided the base & arrested many naval officers & men allegedly involved in the plot. It was reported that the Cienfuegos Naval Base also was raided & that most Cuban naval installations were under army surveillance.

Invasion from U.S. Reported. Cuba's Armed Forces Ministry said Oct. 6 that a 27-man invasion force from the U.S., including 3 Americans, had landed in Oriente Province Oct. 5 between the towns of Baracoa & Moa. It said Cuban militiamen were pursuing the invaders after a brief battle in which the group's leader was killed & 2 invaders captured.

The killed leader was identified as Armentino Feria, a follower of ex-Sen. Rolando Masferrer, who was a Batista supporter. Those captured were Enrique Torres Perez & Arturo Alvarez, an ex-Batista soldier charged with killing 3 persons.

Anthony Zarba, 28, of Somerville, Mass. & 7 Cubans were executed by a firing squad in Santiago de Cuba Oct. 13 after being convicted Oct. 12 by a military tribunal on charges of participating in the Oct. 5 invasion. 18 others received one- to 30-year jail terms. 2 other Americans, Allan D. Thompson, 36, of Queen City, Tex. & Robert O. Fuller, 25, of Miami, were convicted & shot to death Oct. 16 for their alleged part in the invasion. This was the first time Americans had been executed in Cuba since Cuba became a republic in 1912.

A U.S. note Nov. 11 accused Cuba of "discrimination" in executing the 3 Americans. The note, delivered to the Cuban Foreign Ministry by U.S. Chargé D'Affaires Daniel M. Braddock, charged that death sentences were imposed against the 3 Americans because of "their [U.S.] citizenship." It said several Cubans convicted of the same charges were given prison sentences.

The note asserted that no evidence was produced at the trial to show that Zarba was an invasion leader. It said he was denied his "fundamental humanitarian rights" in being refused permission to confer with a lawyer prior to his trial. In the cases of Fuller & Thompson, the note said, there was "wholly inadequate time to prepare an appeal." It said their trials were marked by a "Roman circus atmosphere" & "political harangues bearing no relation to the facts of the case."

Cruz Executed. Roberto Cruz Alfonso was executed by a firing squad

in Havana Oct. 1 after being convicted by a military court on charges of participating in the attempted assassination of anti-U.S. broadcaster José Pardo Llada July 9. 11 others accused in the plot received jail terms ranging from 10 to 30 years Oct. 1, and 6 others were sentenced Oct. 7 to 3-12 year terms.

Executions Mount. 6 Cubans were killed by firing squads in Manicaragua Dec. 10. This brought to 582 the number executed in Cuba since Fidél Castro took power Jan. 1. 1959. 4 of those shot had been convicted Dec. 9 on charges of fatally wounding a Cubana airlines pilot in a gun battle Dec. 8 as they tried to seize his plane as it left Cienfuegos for Miami. A woman passenger, Caridad Cabrera, convicted of taking part in the attempted plane seizure, received a 30-year jail sentence. The 2 others executed had been convicted Dec. 9 after being captured Nov. 30 while fighting as anti-Castro rebels in the Escambray Mountains.

'Conspirators' Jailed. 104 alleged anti-Castro conspirators were convicted by a military court in Havana Feb. 10 and sentenced to prison terms of 3 to 30 years. Ramon Mestre got a 20-year sentence. Ex-Sen. Arturo Hernández Tellaheche got a 9-year sentence. The plotters allegedly had planned to make Mestre premier and Hernández president. 36 defendants were acquitted. All had been arrested in Aug. 1959 in connection with a plot allegedly aided by Dominican dictator Rafael Trujillo to invade Cuba from the Dominican Republic.

Defections. Lt. Cmndr. Miguel F. G. Pons Goizueta, naval attaché at the Cuban Embassy in Washington, resigned Mar. 17 and charged that the Cuban government was "under the most absolute influence of international communism." (Pons Goizueta had headed the Cuban naval delegation to the Inter-American Defense Board & was Cuban delegation adviser to the OAS.) The military & air attaché, Capt. Angel Saavedra, had resigned Feb. 16. As a sergeant attached to the Embassy, he had been a Castro agent prior to the fall of Batista.

The Cuban government charged Mar. 19 that Pons Goizueta had absconded with $110,000. A U.S. district court in Washington issued a restraining order Mar. 19 to prevent further use of an account that Pons had opened in Riggs Bank in his own name with Cuban funds. The account had been opened, he said, on government instructions to shield Cuban purchases of military equipment & other goods. He said he had made only one withdrawal—a $63,000 withdrawal Mar. 11 to settle a bill from N.Y. exporter Eugene Strauss. Pons said he particularly wanted to pay this bill because he had been told that his government was planning to refuse payment on the ground that Strauss "is a Jew, and they said Jews have enough money."

Cuban Amb.-to-Canada Luis A. Baralt, 32, announced his resignation June 16 in Ottawa. He said he disagreed "with my government on fundamental aspects of its policies" and could no longer represent it. Amb.-designate-to-U.S. José Miró Cardona, 58, who had served as first premier in the Castro revolutionary government, resigned July 5 and requested asylum in the Argentine Embassy. He said that "ideological differences" between the government's policies & his own conscience were insoluble. The Foreign Ministry said later July 5 that Miró Cardona had been "removed" from his post as part of a diplomatic service reorganization, as had Amb.-to-Italy Manuel Piedra de la Concha, Amb.-to-El Salvador Francisco Pivadal and Amb.-to-UN (in Geneva) Francisco Garcia Mador.

Dr. Teresa Casuso, a member of the Cuban mission to the UN, announced in New York Oct. 14 that she was resigning and seeking U.S. asylum. She had sent Castro her resignation Oct. 11. Dr. Casuso charged that Castro had turned Cuba into "a police state." But she said Castro was not a Communist and would not lead Cuba to Russia's side.

Lt. José Paz Novas, 42, who said he had headed Cuba's military spy system in the U.S., announced in New York Sept. 26 that he had defected to the U.S. Aug. 20 after he became convinced that the Cuban government was "Communist." Paz said he had been sent to the U.S. in August to command 120 agents, mostly in New York & Miami, who spied on anti-Castro organizations.

He said he had given U.S. officials the names of all spies he could remember. His New York news conference was sponsored by the Cuban Democratic Front, an anti-Castro coalition in the U.S.

Maj. Raúl Chibás, ex-Agriculture Minister, Railroad Commissioner & close associate of Premier Castro, fled Cuba with his wife in a motorboat Aug. 3 & arrived the same day in Key West, Fla. The couple was granted U.S. political asylum.

Chibás, who had been one of Castro's Sierra Maestra guerrillas, said at a news conference in Miami Aug. 4 that he quit Cuba "because that was the only way I could make known to the Cuban people that I did not like what their government is doing." He said Castro dismissed him as Railroad Commissioner in June "because he knew that I didn't like the way they were running things." He said there were many technicians from Communist countries in Cuba & that "Red indoctrination is everywhere."

Emigration Curbed. In a move to prevent technically-trained Cubans from fleeing the country, the government Oct. 31 extended its emigration ban to include engineers, petroleum specialists, government workers & executives of recently-nationalized firms. They were required to receive permits from their employers if they wanted to leave Cuba to visit or to take up foreign residence. Thousands of doctors, lawyers & engineers reportedly had fled the country.

Exiles Plan Castro Ouster. The formation of a Revolutionary Movement of the People to overthrow the regime of Premier Castro was announced in New York Nov. 6 in a manifesto signed by ex-Castro aides who had fled Cuba. The manifesto, made public by Raúl Chibás, said: "Combat [against Castro] has already begun in the mountains & towns of Cuba"; the goals of the Cuban revolution had been "handed over to international communism"; Cuba was in the grip of militarization, internal espionage & "planned terror"; the coffee crop was down 25%; the National Bank had issued 300 million pesos in 11 months to increase currency circulation by 60%; Cuba was fostering subversive expeditions against other governments in the Americas.

Leaflets Urge Revolt. Planes flew over Havana & other Cuban cities Dec. 11-12 and dropped leaflets calling on Cubans to overthrow the Castro government. The leaflets, signed by the Democratic Revolutionary Front, urged Cubans to enlist in an "invisible army of resistance" or to join as "military soldiers" in the mountains.

The coordinator general of the Democratic Revolutionary Front, Dr. Manuel A. de Varona, said in Miami Dec. 13 that the leaflet planes had taken off from outside the U.S.

U.S. Aids Refugees. Pres. Eisenhower, invoking the Mutual Security Act, authorized Dec. 2 the spending of up to $1 million for relief & resettlement of 30,000-40,000 anti-Castro Cuban refugees in New Orleans and in Miami & other Florida areas. The money was to come from the President's $150 million special contingency fund provided by the act's Dirksen-Douglas amendment, which authorized its use "to encourage the hopes . . . of peoples who have been enslaved by communism." Tracy Voorhees was to handle the Cuban refugee problem.

Rep. Francis E. Walter (D., Pa.) told the Council of the Inter-Governmental Committee for European Migration in Geneva Dec. 1 that 36,000 Cubans had been given U.S. asylum & that Cubans were entering the U.S. at the rate of 1,000 a week.

Unionists Get Asylum. At least 12 Electrical Workers Union leaders were granted asylum in the Havana embassies of Argentina, Peru, Ecuador & Brazil Dec. 14. The labor officials fled a few hours before a Havana rally called to approve the "purification" of their union. About 1,000 members of the Electrical Workers' Union demonstrated before the Presidential Palace in Havana Dec. 9 against Communist attempts to seize control of their union. Pres. Osvaldo Dorticós Torrado agreed to meet with a group of 8 electrical union officials headed by the union's secretary general, Amaury Fraginals. After the meeting, Fraginals told the demonstrators not to attend a mass rally scheduled that night by the Communist-controlled Confederation of

Cuban Workers, which sought the removal of Fraginals & his supporters from the electrical union's governing board. Government radio commentators Dec. 9 blamed Fraginals & his supporters for recent bombings in Havana.

Courts Reorganized. The Cuban cabinet Dec. 20 suspended for 45 days a law that barred the removal of judges. Then, by decree Dec. 21, the government ousted 13 judges regarded as disloyal to the Castro regime & reorganized the nation's entire judicial system.

Under the decree, the Supreme Court was to be reformed in 6 weeks by a reduction of the number of judges from 32 to 15 & by its reorganization into a court of 3 chambers (each served by 5 judges) that would handle (1) civil & administrative cases, (2) criminal cases & (3) constitutional, social & agrarian cases. Judicial appointments would be made by the president, premier & cabinet.

Among the 13 judges purged Dec. 21 were Dr. Angel Fernandez Rodriguez, Premier Castro's first justice minister, & 4 Supreme Court judges, including ex-Supreme Court Pres. Emilio Menendez. All of those ousted had fled Cuba.

Newspapers Suppressed. It was announced Mar. 16 that the government had suspended 3 daily Havana newspapers, Excelsior, El Pais & Diario Nacional. Their printing plants were to be converted into a national printing shop for books. The government currently controlled 7 Havana newspapers & most small Cuban newspapers. Havana's Channel 4 TV station was placed under official government management Mar. 26, pending eventual expropriation by the minister of finance. This gave the government control of 5 of the 7 TV stations in the Havana-Camaguey region.

Diaria de la Marina, Cuba's oldest newspaper & critic of the Castro regime, was seized May 11 by a group representing the Collegium of Journalists of Havana & the Printshop Workers Union. Seizure of the paper was denounced May 11 by William H. Cowles, president of the Inter-American Press Association, as "a thinly disguised confiscation." Prensa Libra, a Havana afternoon newspaper, was seized by its workers May 16 after its editor had sought to print an editorial that they considered "insulting" to the government.

The only privately-owned paper, Información, suspended publication Dec. 23. The newspaper reportedly had been losing money because of lack of advertising & government refusal to permit it to reduce its staff. In a front page statement, Información's employes accused the newspaper of a "lockout."

Ammunition Blast. A Cuban army ammunition depot near Havana harbor blew up June 25. 2 persons were killed, 200 injured. An anti-Castro organization in Miami said June 27 that its agents had placed time bombs in trucks parked near the depot.

Castro Vs. Church

'Fascist' Priests Charged. Premier Fidél Castro Aug. 11 accused the Roman Catholic Church in Cuba of "systematic provocations" against the government. Castro, apparently fully recovered after a siege of pneumonia, made the charge at a labor rally in Havana. He said the church had responded to U.S. Embassy & Franco Spain instigation in mobilizing "as many Fascist priests as can be found to fight against the revolution." "Those who condemn" the Cuban revolution "condemn Christ, and they would be capable of crucifying Christ because He did what we are doing," Castro said. In a pastoral letter Aug. 7, members of the Cuban hierarchy had expressed concern at what they called the rise of communism in Cuba & the government's Soviet ties.

Archbishop Assails Reds. A pastoral letter ordered read in Oriente Province's Roman Catholic chuches Sept. 24 charged that Cuba's Communists were trying to destroy all Catholic influence in the country. The letter, signed by Enrique Pérez Serantes, archbishop of Santiago, held that "a few devotees of Marx & Lenin" sought "to take away the well-won laurel of victory" from Cubans who had fought for the revolution and were "ordering us to confine ourselves to temples." The letter denied that the Roman Catholic Church in Cuba was influenced by the U.S. or by the Spanish Falangists. It said the church would not hesitate to choose between North Americans & Russians

but would "always say, 'Cuba yes, communism no; Cuba yes, slavery never.' "

Bishop Attacks Government. Cuban Roman Catholics opposed the "materialistic communism & totalitarianism" of the Castro regime, Asst. Bishop Eduardo Boza Masvidal of Havana asserted in an article in the Catholic magazine La Quincena (distributed in Havana Oct. 30). The article was entitled "Is the Social Revolution of Cuba Christian?" It said the revolution lacked "a spiritual concept of life & man" and "breathe[d] more hate & vengeance than a desire for justice." It criticized Cuba's attacks on the U.S. & deplored its close ties with Communist countries as links that could "only exist between those who think in the same manner & have the same ideals."

Cuban Economic Policy

Business Curbs Tightened. The Cuban cabinet Feb. 20 approved a law creating a Central Planning Board to "supervise & coordinate" the economic affairs of the nation and to determine necessary measures to obtain economic results desired by the government. The board was to be headed by Premier Castro and was to include Maj. Ernesto Guevara, National Bank of Cuba president.

The cabinet Feb. 20 passed measures (a) banning foreign commercial advertisements from movie & TV screens (only ads made in Cuba were to be shown) and (b) requiring that all pharmaceuticals sold in Cuba must be packaged or repackaged in Cuba.

Castro said at a union meeting Feb. 25 that investments of private foreign capital were undesirable because they were made only for profit. Henceforth, foreign investments would be accepted only if delivered to the government to use for "national enterprises" so that the country would not be dependent on the "will of foreign bosses," Castro declared. The new investment rule was made known on the occasion of the first delivery of the 4% deduction from all organized workers' wages. The deduction had been pledged for 5 years to aid industrialization & provide work for 134,000 of an estimated 661,000 unemployed.

Reserves Up. Castro reported in a radio-TV address Apr. 22 that Cuban dollar reserves had risen to $142 million from the low of $77,440,000 at the end of the Batista regime. The increase resulted from import restrictions & new sugar treaties.

Tobacco Industry Seized. The government Sept. 16 seized the entire Cuban tobacco industry (including 16 cigar & 14 cigarette factories and 20 tobacco warehouses) on grounds that the firms were sabotaging Cuba's economic development. The firms were accused of reducing production by 15 million units (nearly 17%) in a year. Among companies seized was the U.S.-owned cigarette firm of Calixto López & Co., valued at about $3 million.

U.S. Agriculture Department officials announced in Washington Sept. 15 that 280 pounds of U.S. tobacco seed that the Cuban Embassy in Washington had tried to smuggle out of the U.S. into Cuba had been seized by U.S. customs agents in New York Sept. 3. The seed, worth $13,440-$18,000, had been mislabeled by the Embassy as barley seed. It could have produced 50 million pounds of cigarette tobacco for world sale in competition with U.S.-grown tobacco. U.S. law restricted its export to an ounce or 2 for research.

Nationalization Decree. Under a decree approved at a cabinet meeting Oct. 13-14, the Cuban government Oct. 14 nationalized about 400 private companies, including 20 U.S.-owned firms valued at $20 million. The seized companies included banks, transportation firms, 105 sugar mills & 61 textile plants.

The U.S.-owned enterprises seized included the Colgate-Palmolive Co.'s Crusellas soap factory, Procter & Gamble's Sabates plant and properties of the Sherwin Williams Paint Co., Pittsburgh Plate Glass Co., du Pont Inter-Americana Chemical Co. and Swift & Co.

U.S. Presidential Campaign

Policy Debated. U.S. policy toward Cuba was debated by U.S. Vice Pres. Richard M. Nixon and Sen. John F. Kennedy (D., Mass.) during the campaign that resulted in Kennedy's election as President. The issue was taken up during a series of Nixon-Kennedy TV debates.

During a TV confrontation in Washington Oct. 7, Nixon rejected a charge made by Kennedy in a Cincinnati TV address Oct. 6 that the Eisenhower Administration was to blame for Fidel Castro's rise in Cuba.

"I don't agree . . . that Cuba is lost," Nixon said. "I believe that we are following the right course, . . .

the only proper [course] . . . which will see that the Cuban people get a chance to realize their aspirations of progress through freedom and that they get that with our cooperation with the other states in the Organization of American States."

Nixon said that "there were 11 dictators in South American & in Central America when we came in in 1953. Today there are only 3 left, including the one in Cuba. We think that's pretty good progress."

Nixon said Kennedy had "indicated . . . that he thought I had made a mistake when I was in Cuba in not calling for free elections" there. Nixon pointed out, however, that in Kennedy's book "The Strategy for Peace," Kennedy had taken "the right position" that the U.S. treaty with all OAS states "prohibits us from interfering in the internal affairs of any other state."

Kennedy retorted that "I never suggested that Cuba was lost except for the present. . . . I indicated that I thought Cuba . . . again would be free." Kennedy said: He had criticized Nixon "because in his press conference in Havana in 1955 he [Nixon] praised the competence & stability of the Batista dictatorship"; Kennedy did not criticize Nixon "for not calling for free elections. What I criticized was the failure of the Administration to use its great influence to persuade the Cuban government to hold free elections. . . . Our aid continued to Batista. . . . We never were on the side of freedom, we never used our influence when we could have used it most effectively—and today Cuba is lost to freedom."

During a TV confrontation in N.Y. Oct. 21, Nixon charged that "Sen. Kennedy's policies & recommendations for the handling of the Castro regime are probably the most dangerously irresponsible recommendations that he has made during the course of this campaign. In effect, what Sen. Kennedy recommends is that the United States government should give help to the exiles & to those within Cuba who oppose the Castro regime—provided they are anti-Batista." Nixon indicated that such action would violate: (a) "5 [U.S.] treaties with Latin America . . . in which we have agreed not to intervene in the internal affairs of any other American country," and (b) the UN Charter's "Preamble, Article I & Article II, [which] also provide that there shall be no intervention by one nation in the internal affairs of another."

Nixon suggested that the U.S. could do in Cuba "what we did with Guatemala. There was a Communist dictatorship that we inherited from the previous Administration. We quarantined Mr. Arbenz [Jacobo Arbenz Guzman]. The result was that the Guatemalan people themselves eventually rose up, and they threw him out. We are quarantining Mr. Castro today. . . . By cutting off trade, by cutting off diplomatic relations, as we have, we will quarantine this regime so that the people of Cuba themselves will take care of Mr. Castro."

Kennedy replied that "if any economic sanctions . . . are going to be successful, they have to be multilateral . . . [and] include the other countries of Latin America. The very minute effect of the action . . . taken this week on Cuba's economy—I believe Castro can replace those markets very easily through Latin America [& Europe]." Kennedy held that if the U.S. had "stronger prestige," it could persuade the countries of Latin America & Europe as well as Canada & Japan "to join in an economic quarantine of Cuba. That's the only way you can bring real pressure."

Kennedy foresaw that "the big struggle will be to prevent the influence of Castro spreading to other countries" of Latin America. He said the U.S. would have "to associate ourselves with the great desire of these people for a better life if we're going to prevent Castro's influence from spreading throughout all of Latin America. His influence is strong enough today to prevent us from getting the other countries of Latin America to join with us in economic quarantine. . . . You, yourself, said, Mr. Vice President, a month ago, that if we had provided the kind of economic aid 5 years ago that we are now providing, we might never have had Castro. Why didn't we?"

Kennedy charged that "today . . . the Russians broadcast 10 times as many programs in Spanish to Latin America as we do. . . . We don't have a single program sponsored by our government to Cuba to tell them our story."

Nixon replied: "We have appropriated 5 times as much for [programs in] Latin America as was appropriated by the previous Administration; . . . we have $2 billion more for the Export-Import Bank; . . . we have a new bank for Latin America alone of a billion dollars"; Pres. Eisenhower won Congressional approval of his new $500 million Latin American program. Kennedy "complains very appropriately about our inadequate radio broadcasts for Latin America," but "his [Kennedy's] Congress—the Democratic Congress—has cut $80 million off the Voice of America appropriations."

Legion Addresses. The Presidential candidates amplified their views in speeches in Miami Beach Oct. 18 before the national convention of the American Legion.

Nixon proposed that Cuba be dealt with by a "quarantine [of] the Castro regime in the Americas." He said "a number of steps . . . are planned," and "we will very promptly take the strongest possible economic measures" against Cuba. Nixon said the Administration had been "patient" with Cuba "because we need & want the support of all of our Latin-American friends . . . for our position. And it [such support] has been developing. . . ."

Kennedy's Cuba Proposals. Kennedy Oct. 20 issued a statement (1) attacking "Mr. Nixon's new Cuba policy" —the economic & diplomatic boycott— as "too little & too late" and (2) proposing a 4-point Kennedy plan for dealing with the Cuban problem.

Kennedy's most controversial proposal was that "we must attempt to strengthen the non-Batista democratic anti-Castro forces in exile, & in Cuba itself, who offer eventual hope of overthrowing Castro." He said that "thus far these fighters for freedom have had virtually no support from our government."

Kennedy also urged that the U.S. (a) work with both the OAS "& our European allies in order to promote collective action against communism in the Caribbean," (b) "consider more stringent economic sanctions—such as the seizing of all Cuban assets in this country to be used to pay off some of the vast sums which Cuba still owes us," and (c) "immediately act to prevent communism from taking over other countries in Latin America—by removing the conditions under which communism thrives."

Kennedy said that "after doing nothing for 6 years while the conditions that give rise to communism grew—after ignoring the repeated warnings of our ambassadors that the Communists were about to take over Cuba—after standing helplessly by while the Russians established a new satellite only 90 miles from American shores—Mr. Nixon &

the Republicans, after 2 years of inaction since Castro took power, now attempt to make up for this incredible history of blunder, inaction, retreat & failure by cutting off several million dollars worth of exports in a move which will have virtually no effect by itself in removing Communist rule from Cuba." "In cutting off a carefully limited number of the American goods which can be sent to Cuba," Kennedy said, "Mr. Nixon has made a dramatic but almost empty gesture . . . which will have so little impact . . . as to be almost meaningless. . . ." He said Castro himself had been cutting American imports to the point where they declined from a value of $42 million in Aug. 1958 to $18 million in Aug. 1960 and that "the goods which are now cut off are the very goods whose export has been declining while the goods which Cuba can still buy . . . are the goods whose export has declined very little."

Nixon charged in a Chester, Pa. speech Oct. 22 that "Sen. Kennedy's call for United States government support of a revolution in Cuba is the most shockingly reckless proposal ever made in our history by a Presidential candidate during a campaign." He added in a Muhlenberg College address later Oct. 22 that Kennedy's proposal violated the U.S.' Latin American & UN Charter commitments and was "a direct invitation for the Soviet Union to intervene militarily on the side of Castro." Such intervention, he said, "can only mean a civil war in which the United States could not avoid entering." He foresaw that the result could "easily" mean "world war."

Kennedy, in a telegram to Nixon Oct. 23, accused the latter of "distortions of the record concerning my position on Cuba." Kennedy said: "You have developed the technique of having your writers rewrite my statements, using these rewritten statements and attacking me for things I have never said or advocated. . . . I have never advocated . . . intervention in Cuba in violation of our treaty obligations. . . . What I have advocated is that we use all available communications . . . & the moral power of the American government to let the forces of freedom in Cuba know that we believe that freedom will again rise in that country."

1961

The U.S. broke diplomatic relations with Cuba's Castro regime early in 1961 and backed Cuban exiles in an unsuccessful invasion of Cuba. Fidel Castro announced that he was leading Cuba to communism.

U.S. Pres. Kennedy started a sweeping Alliance for Progress to bring economic, social and political betterment to Latin America.

Fidel Castro

Wide World

U.S. Ends Relations With Cuba

Relations between Cuba and the U.S. deteriorated during 1960. In 1961 the U.S. ended direct diplomatic contacts with Cuba. Among factors that led to the break: Castro's growing ties to the Soviet bloc; U.S. fears of Castro-fomented uprisings throughout Latin America; Castro's "vilification" of the U.S.; Castro's seizure of U.S. assets in Cuba; Castro's attacks on U.S. rights to the Guantanamo naval base; the U.S.' refusal to renew its Cuban sugar-purchase contracts; Cuban fears of a U.S.-based invasion.

Diplomatic Break. The U.S. severed diplomatic relations with Cuba Jan. 3. Announcing the break in ties, Pres. Dwight D. Eisenhower said in a White House statement: "There is a limit to what the United States in self-respect can endure. That limit has now been reached."

The rupture, after more than a year of increasing antagonism, was precipitated by Cuban Premier Fidel Castro's Jan. 2 demand that the U.S. reduce its Havana embassy staff from 87 Americans to 11 within 48 hours. Castro charged that 80% of the embassy's 300 employes were FBI and Pentagon spies. He accused the embassy of "paying . . . [Cuban] terrorists to put bombs in Cuba."

In his Jan. 3 statement, Mr. Eisenhower said Cuba's "unusual" demand "can have no other purpose than to render impossible . . . normal [U.S.-Cuban] diplomatic relations." He termed it a "calculated action" that was "the latest of a long series of harassments, baseless accusations & vilifications." The President, expressing hope for resumption of U.S.-Cuban relations "in the not-too-distant future," voiced "sympathy" for Cubans under "a dictator."

Formal notification of the diplomatic break was made in a note handed by U.S. State Secy. Christian A. Herter Jan. 3 to the Cuban chargé in Washington, Dr. Armando Florez-Ibarra. The Swiss government Jan. 4 agreed to Herter's request that it represent U.S. diplomatic interests in Cuba. Swiss Amb.-to-Cuba Walter Bossi assumed the responsibility Jan. 7. The Czech government announced

33

Jan. 4 that it would handle Cuba's diplomatic affairs in the U.S.

The Jan. 3 U.S. note asked Cuba to withdraw its 101 staff members from its Washington embassy and its 15 consular offices in the U.S. "as soon as possible," but it set no time limits.

Herter Jan. 3 asked State Secy.-designate Dean Rusk whether the incoming Kennedy Administration wanted to associate itself with Mr. Eisenhower's decision. After talks with Pres.-elect John F. Kennedy, Rusk told Herter that the new Administration could not commit itself.

The Cuban government Jan. 5 blamed the Eisenhower Administration for the break. A communiqué issued after a Cuban cabinet meeting said the U.S. first caused "the deterioration of these relations with its aggressive and unfriendly policy against" Cuba. It defended Cuba's decision ordering the U.S. to reduce its embassy staff on the grounds that embassy officials had "furthered counter-revolution and terrorism."

International Reaction. The Soviet news agency Tass Jan. 4 condemned the U.S. action as a "plot to commit aggression against the revolutionary Cuban government." Communist China Jan. 4 called the U.S. decision a prelude "to overthrow [of] the Cuban government." The Yugoslav government criticized the U.S. for the break Jan. 6.

Latin-American reaction was mixed: Government leaders in Haiti, Peru, Guatemala and Colombia endorsed the U.S. action Jan. 4; Ecuadorian Foreign Min. José I. Chiriboga expressed "deep sorrow"; leftist and Communist newspapers in Chile criticized the U.S., but the Chilean government was noncommittal, as were the governments of Argentina, Venezuela, Brazil and Bolivia.

(6 Latin American nations had previously broken relations with Cuba: the Dominican Republic, Paraguay, Nicaragua, Haiti, Guatemala and Peru.)

U.S. Aides Leave. U.S. officials Jan. 4 started preparations to close the U.S. embassy in Havana, and the staff began leaving Havana the same day when 63 embassy employes sailed to West Palm Beach, Fla.

The State Department disclosed Jan. 3 that it had urged the 3,000-3,500 American civilians remaining in Cuba to leave. A department spokesman said that, if necessary, the U.S. might help in evacuating them.

UN Rejects Invasion Charge. The UN Security Council Jan. 5 rejected without a vote Cuban charges Jan. 4-5 that the U.S. planned an "immediate" invasion of Cuba and that Guatemala, Honduras and Florida had camps for training "mercenaries." The Council adjourned after hearing Cuban Foreign Min. Raúl Roa demand that the U.S. be labelled an aggressor for moves that, he claimed, had led to the severance of U.S.-Cuban relations.

U.S. Amb.-to-UN James J. Wadsworth said in Council debate Jan. 4 that the U.S. broke relations with Cuba because the Castro government had made normal diplomatic contacts "virtually impossible." Wadsworth denied that the U.S. planned to invade Cuba.

Cuban Moves Vs. 'Invasion.' The Cuban government Jan. 6 took military precautions against a "Yankee invasion" that Castro had predicted would occur before the Kennedy Administration took office Jan. 20. Exit permits for Cubans and foreign residents, except Americans, were suspended. Thousands of troops were called out.

The Havana newspaper Revolución Jan. 7 reported the seizure of 6 "terrorists" and of 2 air-dropped cargoes of arms that had been sent to rebels by "agents of the [U.S.] Central

Intelligence Agency." The ammunition was said to have been dropped in Pinar del Rio and Escambray Provinces.

The Cuban army Jan. 8 reported the arrest of 10 alleged terrorists in the Havana suburb of Cotorro. They were said to be members of an underground group directed by exiled ex-Public Works Min. Manuel (Manolo) Ray. The government press said the men were seized with U.S.-made arms "furnished by imperialists." Thousands of other arrests were reported.

Castro for Resuming Tie. Castro declared Jan. 20 that Cuba was ready to "begin anew" its diplomatic relations with the U.S. provided that the U.S. initiated the improvement of relations. Castro said he saw in the Kennedy Administration's assumption of office "a little hope . . . for peace."

Castro made this overture at a rally of 50,000 armed civilian militiamen who were being demobilized after the call-up to block the "Yankee invasion." He said the invasion threat had ended when the Eisenhower Administration left office.

Castro Feb. 14 repeated his hopes for renewing U.S.-Cuban diplomatic relations, but only if Pres. Kennedy ended the alleged U.S. policy of interfering in Cuba's internal affairs. Castro said that "direct [U.S.] military support" of anti-Castro rebels in the Escambray Mountains of Cuba impeded "peaceful relations."

(A U.S. policy statement Feb. 24 held that U.S.-Cuban disputes could be settled only if the Castro régime "establish[ed] freedom . . . so that [Cubans] may freely choose their destiny." The statement, made by State Department spokesman Lincoln White, said the U.S. was sympathetic with the aims of the Cuban revolution but that it was concerned "with the capture of the revolution by external influences for the purposes of impos-

ing on Cuba an anti-democratic ideology . . . and exporting that ideology by subversion to other American republics.")

Mediation Offers Rejected. Argentine Foreign Min. Diógenes Taboada proposed in messages to the U.S. and Cuba Mar. 4 that Argentina mediate the dispute between the 2 countries. The Cuban government Mar. 8 rejected the offer on the ground that the "conduct of the [U.S.] government and public declarations of its officials" showed that the U.S. was not ready to negotiate.

A Mar. 22 note rejecting the mediation offer, signed by U.S. State Secy. Dean Rusk and delivered to Taboada by U.S. Amb. Roy R. Rubottom Jr., said the U.S. would discuss the Cuban dispute if Cuba fulfilled its inter-American obligations and dissociated itself from the imperialistic designs of the USSR and Communist China.

U.S. Retains Guantánamo. A White House statement Jan. 4 said that the severance of U.S.-Cuban relations would have no effect on the status of the U.S.' Guantánamo naval base. It added that the 1903 and 1934 U.S.-Cuban treaties granting the U.S. the right to the base could not be abrogated without U.S. consent.

Revolución charged Jan. 9 that the U.S. was mining Guantánamo Bay. It said the U.S. aircraft carrier Franklin D. Roosevelt, which sailed into Guantánamo the same day, was part of a fleet the U.S. planned to station in the area. The U.S. said that the carrier, which stayed in port 5 hours to unload replacements, was to take part in maneuvers off Puerto Rico.

Cuban Pres. Osvaldo Dorticós Torrado said at a Havana rally Jan. 13 that Cuba would claim the Guantánamo base "in proper time." He said, however, that Cuba would not attack the base despite U.S. "provocation."

The U.S.' right to the Guantánamo base was challenged Aug. 24 by Mario Garcia-Inchaustegui, chief Cuban delegate to the UN. Speaking before the UN General Assembly, Garcia denounced the 1903 and 1934 treaties. He said: "Neither under law nor under international morality can any validity . . . be given to documents signed for the purpose of diminishing the sovereignty of states"; nations "cannot be sovereign if they are forced to admit foreign military forces in their territory"; "mercenaries are being concentrated at Guantánamo and in Puerto Rico to be used later against small independent states."

U.S. Amb.-to-UN Adlai E. Stevenson, terming Garcia's challenge to the Guantánamo treaty "an extraordinary new doctrine of international . . . lawlessness," called it "a normal exercise of sovereignty" for nations to negotiate defense agreements.

A Cuban government decree announced Jan. 19 required all of the 3,000-3,500 Cubans working at the Guantánamo base to obtain Cuban government permits before entering the base. The Cuban government Feb. 4 seized the Yateras Water Co. aqueduct that supplied water to the base. The base's water supply was uninterrupted, however. The government said it had seized the water firm because its owners, members of the Bacardi and Schueg families, had fled Cuba.

Cuban Rebel & Exile Activity

Miró Heads Exiles. The N.Y. Times reported Mar. 22 that 2 rival anti-Castro groups in the U.S.—the Democratic Revolutionary Front, headed by Dr. Manuel Antonio de Varona, and the Revolutionary Movement of the People, headed by Manuel Ray—had met in Miami Mar. 20 and selected José Miró Cardona, 59, Cuban premier

during the early days of the Castro régime, as president of an anti-Castro National Revolutionary Council in the U.S. The council eventually was to become a "government in arms" with the purpose of establishing a military foothold in Cuba for the overthrow of the Castro régime.

Negotiators of the agreement to select Miró as president: for the Democratic Revolutionary Front, Dr. Justo Carillo Hernández, former member of the World Bank board of governors and ex-president of the Cuban Bank for Agricultural & Industrial Development, Manuel Artime and ex-Treasury Min. José Alvarez Díaz; for the Revolutionary Movement of the People, Maj. Raúl Chibás, Ray, Armando Loré, and Dr. Félipe Pazos, ex-Cuban Bank president.

The front was said to represent a middle-of-the-road philosophy. Miró's selection was formally announced at a news conference in New York Mar. 22. The 2 organizations said they had agreed on this program: (a) general elections within "18 months following the overthrow of the Communist tyranny" (the provisional president would be barred from running in the election); (b) return of confiscated assets except those regarded as necessary for the national interest (owners of the latter would receive proper reimbursement); (c) agrarian reform; (d) elimination of "state control of labor"; (e) low cost housing; (f) free enterprise and private ownership; (g) retention and promotion of armed forces members who helped overthrow the Castro régime; (h) declaring the Communist Party illegal; (i) renunciation of foreign treaties threatening the security of the Americas.

(At a rebel organizational meeting in Washington Mar. 10 a spokesman for the exiles said that Castro and his aides would be permitted to leave

Cuba unharmed after the island's "liberation." The exiles pledged that no executions would occur.)

U.S. State Department spokesman Lincoln White said in Washington Mar. 22 that the establishment of a Cuban provisional government in the U.S. without U.S. permission would violate U.S. sovereignty and international law. White said "no such consent has been asked [by the Cuban refugees] or given."

Miró Calls for Revolt. Miró issued in New York Apr. 8 a council statement calling on Cubans to use armed force to overthrow the Castro régime.

Miró charged that Castro's policies had resulted in the channeling of "all [Cuban] resources" into "military equipment, propaganda and instruments of repression" to "sustain a tyranny" and "to export revolution" to other countries in the Western Hemisphere. He said Castro had betrayed the Cuban revolution by: (a) "abolishing the freedom of unions"; (b) denying peasants title to their land; (c) "crushing freedom of expression"; (d) "converting [schools] into centers of communism"; (e) "making it impossible for [judges] to exercise their sacred duty of administering justice."

Miró said at a New York news conference Apr. 7 that Castro would be overthrown by a revolution, not an invasion.

Cuban Invasion

Rebels Land on Beach. An anti-Castro force landed on the swampy beaches of Cuba's southern Las Villas Province before dawn Apr. 17 in an attack directed by the U.S.-based National Revolutionary Council.

Cuban Premier Fidel Castro declared a state of national alert Apr. 17 and called all militiamen to duty.

Revolutionary Council Pres. José Miró Cardona, in a statement issued at council headquarters in New York

Apr. 17, called on "freedom-loving peoples" of the Americas for moral and material aid to the rebels.

The invasion's field operations were led by Capt. Manuel Artime, 29, an ex-Castro officer who had defected in 1960 because, he said, he believed Castro was a Communist.

The rebels, supported by planes, made their principal landing in the area of Bahía de Cochinos (Bay of Pigs) in southern Las Villas Province and penetrated into Matanzas Province, 10 miles from the beachhead. Some rebel paratroopers were dropped further inland. Smaller groups of rebels reportedly landed near Santiago de Cuba in Oriente Province and in Pinar del Rio Province.

A strong government counter-attack was under way by Apr. 18. With government success virtually assured Apr. 19, the Revolutionary Council asserted in a communiqué that the landings had been "inaccurately" described as an invasion. It said the main purpose of the invaders, who "numbered in the hundreds, not the thousands," was to bring ashore "supplies and support for our patriots who have been fighting in Cuba for months."

Invaders Defeated. Castro announced in a communiqué Apr. 20 that government troops had captured "the last points held by the foreign mercenary invasion forces." He said that the Las Villas beachhead had been destroyed and that Playa Giron, in the original Apr. 17 landing area at Bahía de Cochinos, was "the last mercenary stronghold" and was taken by his forces Apr. 19.

Castro's communiqué said that Cuban planes had bombed and sunk several ships in which invaders had tried to escape. It said "a large quantity of arms manufactured in the U.S. were captured, including several Sherman heavy tanks."

Cuban strategic points, including Cochinos Bay (Bahía de Cochinos or Bay of Pigs) invasion area, are shown in Wide World map.

The communiqué admitted that defending Cuban troops, estimated at 20,000 men, "had to pay a high price in precious lives." In a radio-TV address Apr. 23, Castro said 87 Cuban soldiers had been killed and 250 wounded.

In a communiqué issued Apr. 21 the Revolutionary Council admitted that the rebel landing "did not achieve all of its proposed objectives." It said, however, that "the majority of the combatants succeeded in moving inland, . . . where they continue to fight oppressive communism." Rebel leaders asserted the Bahía de Cochinos landing was achieved by 500 men with the support of 5 tanks.

Artime and 21 other rebels were seized by government troops May 2 at San Blas in the Zapata swamp of Las Villas Province. The government said their capture brought to 1,122 the number of rebels taken prisoner.

Havana radio reported May 3 that Artime told his captors that the U.S. had organized, directed and financed the rebel force. Artime was quoted as saying U.S. destroyers had escorted the rebels from Puerto Cabezas, Nicaragua to Cuba.

Invasion Preparations. It was widely reported that the Eisenhower Administration had planned the invasion and that the training of Cuban exiles in Guatemala, under the guidance of CIA Deputy Director Richard M. Bissell Jr., was well under way when John F. Kennedy became President.

The N.Y. Times reported Apr. 21 that Mr. Kennedy had decided against supporting a large-scale invasion of Cuba. Instead, he was said to have agreed to provide ships and other equipment to help the Cuban rebels make smaller-scale landings for the purpose of reinforcing anti-Castro rebels already on the island.

The N.Y. Times Apr. 22 described these pre-invasion developments:

■ The Cuban guerrillas had been trained since May 1960 at 7 camps in Guatemala* under the supervision of a U.S. Army colonel. At least 7 U.S. pilots trained Cuban flyers at a Guatemalan airstrip at Retalhuleu. Some Cuban officers were sent to the U.S. Army's Jungle Warfare School in Panama.

■ The guerrillas, joined by others from Florida and Louisiana camps, assembled early in April at Caribbean islands, including Great Corn and Little Corn islands (owned by Nicaragua and leased to the U.S.), and embarked for Cuba.

■ Events just prior to the invasion were marked by political friction among the Cuban exile groups and by Cuban rebel opposition to the strategy of the U.S.' Central Intelligence Agency. The Revolutionary Council ordered the expulsion from the guerrilla forces of former followers of ousted Cuban dictator Fulgencio Batista. The CIA reportedly opposed the political dismissal of trained men. Most Cubans in the Miami area, including many Castro agents, were said to have been aware by early April that an invasion was imminent.

(The N. Y. Times had reported Jan. 10 that foreign military experts, "mostly" from the U.S., were training a guerrilla force near Retalhuleu. Political foes of Guatemalan Pres. Miguel Ydígoras Fuentes charged that the military activity was for an invasion

*The Times Apr. 22 listed the locations of the Guatemalan camps as La Suiza, Helvetia, Trax, Champerico, Retalhuleu, San José Buenavisea, and Sayaxche.

Rebels who landed Apr. 17 reportedly had trained at Retalhuleu and 6 other Guatemalan camps (1) before boarding invasion vessels at Puerto Cabezas, Nicaragua (2), at Nicaraguan off-shore islands of Great Corn and Little Corn and at other Caribbean points.   (Wide World map)

of Cuba. Ydígoras, denying aggressive plans, said Jan. 10 that U.S. officers were training Guatemalan troops in the Retalhuleu area to repel a possible attack from Cuba. Ydígoras confirmed Jan. 10 that an air base 3 miles west of Retalhuleu was being used for military purposes.)

The N. Y. Times said Apr. 22 that rebels in Miami were bitterly criticizing the CIA for precipitating the Cuban landings without coordinating the attack with Cuban underground sabotage. The rebels also reportedly criticized the CIA for favoring the Council's Movement of the Revolutionary Recovery, led by Artime, over the council's People's Revolutionary Movement, led by Ray, which the CIA reportedly had regarded as too radical.

A pre-invasion bombing attack on 3 Cuban air bases had been carried out Apr. 15 by 3 U.S.-made B-26 bombers piloted by Cubans. They were said to have bombed Camp Libertad outside Havana, at Santiago de Cuba and at San Antonio de Los Baños, 25 miles from Havana. 7 militiamen were killed and 39 wounded in the Camp Libertad attack.

One of the attacking planes reportedly crashed into the ocean off Santa Fe. 2 planes landed in Florida—one, with 2 men aboard, coming down at Key West and the other, with only the pilot aboard, landing in Miami. The 3 Cubans received U.S. asylum.

The Cuban airmen asserted that they had defected with the 3 bombers flown in the attack. Cuban Pres. Osvaldo Dorticós Torrado denied that the airmen were defectors and charged that the attacking planes had come from the U.S.

The picture of one pilot was identified, according to the N.Y. Times, as that of a Lt. Zuniga of the Retalhuleu trainees. The Apr. 28 issue of Time magazine reported that the bombings had been carried out by B-26s flown from bases outside Cuba. It reported that an old Cuban B-26, apparently that flown by Zuniga, had been machinegunned on the ground in the U.S. to make it appear it had suffered damage in the raid on Cuba. It said the plane then was flown to Miami, where the pilot claimed to have defected with the plane and to have carried out attacks in Cuba en route to the U.S.

Invasion Aftermath

International Reaction. The landings touched off anti-U.S. riots and pro-Castro demonstrations throughout Latin America and Europe Apr. 17-22.

Argentinians Apr. 17 smashed windows at the U.S. Information Service building in Buenos Aires. Colombian police halted a pro-Castro mob marching on the U.S.' Bogotá embassy, and pro-Castro demonstrations were reported Apr. 17 in Caracas, Venezuela; Montevideo, Uruguay, and Santiago, Chile.

A resolution condemning "foreign armed intervention" in Cuba was unanimously approved Apr. 17 by the Venezuelan Chamber of Deputies.

The largest Latin-American demonstrations occurred Apr. 21 in Mexico City. About 1,000 soldiers, police and firemen intercepted 15,000 pro-Castro sympathizers who were marching on the National Palace. 150 demonstrators were injured, and 200 were arrested. More than 8,000 pro-Castro Uruguayans demonstrated in Montevideo Apr. 21.

Thousands of persons demonstrated before the U.S. embassy in Moscow Apr. 18. The demonstrators, fought by soldiers and police, carried anti-American slogans and hurled rocks and ink bottles at the building, shattering windows and spattering walls. (The Soviet embassy in Washington was the target of egg-throwing anti-Castro pickets Apr. 20.)

Anti-U.S. demonstrations occurred Apr. 18 in front of U.S. embassies in Warsaw, Paris and Tokyo and before the U.S. legation in Sofia, Bulgaria. The U.S. Information Service building in Belgrade, Yugoslavia was attacked.

Yugoslav Pres. Tito, on a visit to the UAR, said in Alexandria, Egypt Apr. 18 that the Cuban invasion was "not only an attack on Cuba . . . but also on peace. . . ."

In a message to Castro, UAR Pres. Gamal Abdel Nasser called the invasion "a crime against the Cuban people, who are . . . trying to insure genuine freedom for themselves. . . ."

Indian Prime Min. Jawaharlal Nehru, in a statement to Parliament Apr. 20, said he could not "see how it [the Cuban rebel landing] could take place without the organization, encouragement and help of the authorities—public or private—of the United States."

Khrushchev-Kennedy Exchange. Soviet Premier Khrushchev, in a message sent to Pres. Kennedy Apr. 18, charged that the "armed bands which invaded [Cuba] . . . had been trained, equipped and armed" in the U.S. He also blamed the U.S. for the Apr. 15 bombings of Cuban air bases. Khrushchev urged Mr. Kennedy "to call a halt to the aggression against" Cuba. He warned that the USSR would "render [Cuba] . . . all necessary assistance in beating back the armed attack." Khrushchev hinted that the Cuban affair might "lead the world to a military catastrophe."

In a reply to Khrushchev the same day, Mr. Kennedy asserted that the U.S. "intends no military intervention in Cuba." He warned, however, that the U.S. would "honor its obligations under the inter-American system" in case of "military intervention by an outside force." Mr. Kennedy said the invasion of Cuba represented a "struggle" by people who had "found intolerable the denial of democratic liberties and the subversion [of their revolution] . . . by an alien-dominated régime."

(Mr. Kennedy had said at his news conference Apr. 12: U.S. armed forces would not be used to help overthrow Castro's régime; the U.S. would try to "make sure" that no Americans participated in anti-Castro activities in Cuba: he would oppose any attempts by Cuban exiles to mount an offensive against Castro from the U.S.)

Khrushchev, replying Apr. 22 to Pres. Kennedy's note, repeated his charge that the U.S. had "prepared" and "financed" the Cuban invasion and "transported the . . . mercenaries [who] invaded" the island. He said U.S. armed forces had taken "a direct part . . . in the gangster attack.

U.S. National Unity Sought. Pres. Kennedy conferred with ex-Pres. Eisenhower Apr. 22 and with ex-Vice Pres. Richard Nixon Apr. 20 in an effort to unify Americans in the face of the Cuban crisis.

The Apr. 22 meeting was held at the Presidential Camp David, Md. retreat at the request of Mr. Kennedy, who said he had sought "the benefit of his [Eisenhower's] thoughts and experiences." When asked by a newsman whether he supported Mr. Kennedy's stand on Cuba, Eisenhower replied that the U.S. should support "the man who has to carry the responsibility for our foreign affairs."

Mr. Kennedy conferred with Nixon at the White House in Washington. (Nixon said in New York Apr 21 that he had told the President that he would support him "even to commitment to American armed forces" to "stop the buildup of the Communist beachhead in Cuba.")

The President also visited ex-Pres. Herbert Hoover and Gen. of the Army Douglas MacArthur in New York Apr. 28, reportedly to discuss Cuba.

Kennedy Hints 'Intervention.' Pres. Kennedy hinted Apr. 20 that the U.S. might take the step of "intervention" in Cuba if circumstances warranted. Delivering a radio-TV speech at the annual meeting of the American Society of Newspaper Editors in Washington, Mr. Kennedy asserted that "we do not intend to abandon" Cuba to the Communists.

"Should it ever appear that the inter-American doctrine of noninterference merely conceals or excuses a policy of non-action," Mr. Kennedy said, "if the nations of this hemisphere should fail to meet their commitments against outside Communist penetration, then I want it clearly understood that this government will not hesitate in meeting its primary obligations, which are the security of our nation. Should that time ever come, we do not intend to be lectured on intervention by those whose character was stamped for all time on the bloody streets of Budapest."

Denying Castro's charges that the rebels who made the landing were "mercenaries," Mr. Kennedy described them as "a small band of freedom fighters" who had "engaged the armor of totalitarianism." He said Americans were "not complacent about Iron Curtain tanks and planes less than 90 miles" from the U.S. But, he asserted, Cuba was "less a threat to our survival than it is a base for subverting the survival of other free nations" in the Western Hemisphere.

UN Debate. At an emergency meeting of the UN General Assembly's Political Committee, called Apr. 15 before the rebel invasion, Cuban Foreign Min. Raúl Roa charged that the air attacks against Cuban air bases that day were the "prologue to the large-scale invasion" that he said was being planned by the U.S. and other Latin American countries.

Roa charged in a committee speech Apr. 17, the day of the rebel landings, that the U.S. had "organized, financed and armed" the landing force, "which had come from Guatemala and Florida." "These crimes and depredations have been sanctified, paid for and blessed by the U.S. State Department, the Pentagon and the CIA," Roa said. He charged that forces from the U.S.' Guantánamo naval base had entered Oriente Province and had clashed with Cuban troops.

U.S. Amb.-to-UN Adlai E. Stevenson, replying Apr. 17, called Roa's charges "totally false." Stevenson said "no offensive has been launched from Florida" or any other part of the U.S. and that no U.S. forces were fighting in Cuba. (Stevenson said that "'in the last few months" Cuba had received large quantities "of Soviet [military] equipment," including "at least 15 Soviet 50-ton tanks, 19 Soviet assault guns, 15 Soviet 35-ton tanks, 78 Soviet 76-mm. field guns, 4 Soviet 122-mm. field guns and over 100 Soviet heavy machineguns." He said that "over 200 Soviet and Czechoslovak military advisers are in Cuba, and over 150 Cuban military personnel have been sent to the [Soviet] bloc for training.")

Miró Denies U.S. Aid. Revolutionary Council Pres. Miró Cardona insisted at a New York news conference Apr. 21 that the U.S. government had not given his anti-Castro group military or any other kind of help in its rebellion. He said the council did not plan to ask the U.S. for aid. Declaring that "this is a Cuban fight," Miró said the council would "graciously accept" "moral support" from Western Hemisphere nations.

U.S. Senate Probes. State Secy. Rusk testified May 1 at a closed hearing of the U.S. Senate Foreign Relations Subcommittee on Latin American Affairs. Subcommittee Chrmn. Wayne L. Morse (D., Ore.) afterwards quoted Rusk as confirming that the U.S. had trained, financed and equipped the Cuban invaders. Rusk was said to have testified that all U.S. government agencies involved had unanimously recommended the attack. Morse quoted Rusk as saying the U.S. had no current plans "to proceed in any way in armed intervention in Cuba." Rusk asserted, however, that the U.S. would "defend itself" if Cuba committed aggression, such as an attack against the U.S. naval base at Guantánamo. Morse said he was satisfied that the Cuban operation had been a "joint operation" of the White House, the State and Defense Departments and the CIA.

Morse said in a TV broadcast in Washington May 14 that no witness had given his subcommittee "a bit of evidence" to support charges that the Cubans had used Soviet MiGs to repel the invaders or even "that there was a single Russian MiG in Cuba." Morse said: "The planes that were used against the planes of the Cuban exiles were American planes that Castro got possession of as a result of our [the U.S.'] first making them available to Batista."

After a closed subcommittee hearing May 19, Sen. Albert Gore (D., Tenn.) urged the replacement of all members of the U.S.' Joint Chiefs of Staff for their part in the invasion. He did so after hearing testimony by the Joint Chiefs' chairman, Gen. Lyman Lemnitzer. By contrast, Morse expressed "complete confidence in the Joint Chiefs of Staff."

Sen. Frank Church (D., Ida.), also a subcommittee member, said May 19 that the CIA and the Defense Department shared the blame for the invasion's failure He said both agencies had "advised the President that the [invasion] plan was feasible."

CIA Director Allen W. Dulles and Richard M. Bissell Jr., his deputy, who had been directly in charge of the rebel operation, testified May 2 at a secret session of the Senate Foreign Relations Committee. After the meeting, Committee Chrmn. J. W. Fulbright (D., Ark.) called the Cuban attack a "mistake" and said all U.S. agencies involved bore "collective responsibility" for it.

(Dulles was succeeded as CIA director Sept. 27 by ex-Atomic Energy Commission Chrmn. John A. McCone.)

Kennedy Accepts Blame. White House Press Secy. Pierre Salinger Apr. 24 had issued a statement saying that Pres. Kennedy assumed full responsibility for the U.S.' part in the Cuban attack. Salinger said: Pres. Kennedy had restated his previous position "that as President he bears sole responsibility for the events of the past days. . . . The President is strongly opposed to anyone . . . attempting to shift the responsibility."

Interior Secy. Stewart L. Udall had charged in a TV interview Apr. 23 that the Eisenhower Administration had started the steps toward the attack "and handed it over. Eisenhower directed it. Another Administration carried it out." But he said Apr. 24 that he had not intended the impression given by his TV remarks. "The President and his Administration take full responsibility in regard to our relations with Cuba," Udall said.

U.S. Rep. William E. Miller (R., N.Y.), GOP national chairman, charged in a TV-radio broadcast June 11 that the invasion had failed because Pres. Kennedy had "rescinded and revoked the Eisenhower plan to have the Cuban freedom fighters protected by American airpower." Miller said Mr. Kennedy's "most tragic" mistake was his announcement "to the world that he had rescinded this order, that the Americans wouldn't do a single thing to help the Cuban . . . fighters."

Sen. Hugh D. Scott (R., Pa.), appearing on the same program, attributed the failure of the invasion to Mr. Kennedy's refusal to "support th. Cubans with American naval forces, American supplies and American air cover."

Sen. Joseph S. Clark (D., Pa.), also on the program, said: "The whole [U.S.] Cuban policy resulted from a plan conceived by Mr. Eisenhower, known in full by Mr. [ex-Vice Pres. Richard] Nixon, although the knowledge was denied by him during the [1960 election] campaign, which called for the violation of our treaties, the violation of our agreement" with the OAS and the invasion of Cuba with U.S. troops; "Pres. Kennedy had the wisdom not to go in with those American troops."

Eisenhower's Rôle. Ex-Pres. Eisenhower said at a Gettysburg, Pa. news conference May 1 that he was opposed to "a full 'witchhunting' investigation" of the Cuban invasion. Instead, he advised the U.S. to support Pres. Kennedy's "effort to prevent the solidifying of a Communist stronghold" in the Western Hemisphere. The ex-President said that his Administration had provided "some training and equipment" to Cuban exiles but had made no further plans because the exiles lacked effective leadership.

Eisenhower said at a Cincinnati news conference June 12 that he had ordered Mar. 17, 1960 that Cuban refugees in the U.S. be organized, trained and equipped for possible action. He denied, however, that he had forwarded a specific plan to provide U.S. air cover for a Cuban invasion since no exact invasion plan had been drawn up. Eisenhower said he had advised that "no amphibious operation could succeed unless it had air cover."

Eisenhower asserted in Gettysburg Sept. 11 that "there was absolutely no planning for an invasion" of Cuba during his administration. He said he had agreed in Mar. 1960 to help train anti-Castro Cubans in the U.S. But lack of cohesion among the Cuban exiles, he said, would have made it impossible to carry out more ambitious plans.

A White House source reportedly declared Sept. 11: Eisenhower "was in error"; "I not only know there were plans for an invasion while he [Eisenhower] was in office; there are documents to prove it."

Rebel Force Disbanded. The N.Y. Times reported Apr. 27 that the Revolutionary Council's anti-Castro force was no longer in existence. "Reliable sources" in Miami were cited as saying the force had barely exceeded 2,000 men although 4,000-5,000 had been claimed.

The Times said: Of the 2,000 men, 300 were weeded out at guerrilla training camps before the landings and 1,500 landed at Bahía de Cochinos.° Most of those in the landing were either killed or captured. Many were reportedly drowned when 2 invasion vessels were sunk by Cuban planes. But some made their way into the Escambray Mountains to join guerrillas already there. The whereabouts of another rebel force of 200 men, trained in Louisiana and led by Capt. Higinio (Nina) Díaz, remained unknown. The force had been scheduled to land at Morón on the northern shore of Camaguey Province, but Díaz, who learned of the order at sea, reportedly had refused to make the landing because he was said to be unfamiliar with the area.

New Invasion Plans Reported. Sen. George A. Smathers (D., Fla.) said June 25 that he had been told that Cuban exiles were training in Florida for another invasion of Cuba.

The Miami Herald reported June 27 that a group of 50 volunteers, including 25 Americans, known as the Intercontinental Penetration Force, was training in Miami. The Herald said that the group's leader, Jerry Patrick, had stated officially that the force was taking a civil defense course but privately admitted that his group was practicing for possible raids on Cuba.

The Herald called the force part of the International Anti-Communist Brigade of the Junta Revolucionaria de Liberación Nacional, a Cuban exile group headed by Aureliano Sánchez Arango. (Sanchez had proposed in New York Apr. 26 a sweeping reorganization of rebel plans to over-

---

°The U.S. government July 31 ended a semi-secret program of providing direct pay (believed to be through the CIA) to approximately 1,300 Cuban rebels who were not captured in the landing. Each got a $100 interim relief allowance for August and then was made eligible for relief through the U.S. Health-Education-Welfare Department's Cuban Refugee Emergency Center.

throw Castro. In a memorandum said to represent the views of 61 Cuban rebel organizations, the junta criticized the National Revolutionary Council as being under "complete subordination to a U.S. agency" [the CIA].)

Pres. Kennedy had said at a Washington news conference May 5 that the U.S. was "not now training and . . . [was] not now planning to train a Cuban [exile] force" in the U.S. or elsewhere to help overthrow Castro's régime.

Carlos Prío Socarras, 58, ex-Cuban president who had been overthrown by Fulgencio Batista in a 1952 coup, announced in Miami Sept. 2 that he would work with U.S. Rep. Victor L. Anfuso (D., N.Y.) in organizing a Cuban government-in-exile.

Ray Group Quits Council. The People's Revolutionary Movement, led by Manuel Ray, withdrew from the National Revolutionary Council following a meeting in Miami May 22. The Ray movement, which had an active underground in Cuba, was disturbed over what it called CIA approval of council recruitment of ex-Batista followers for a new "national army" in preference to Ray's group, which had been barred from the Apr. 17 invasion.

Ray told newsmen in Miami May 27 that the council had failed to fulfill 3 prerequisites his group had stipulated for joining the council Mar. 22. The conditions, as incorporated in a signed document Mar. 23, were that: (1) "The council . . . give first priority to the fighters who are already inside Cuba fighting the Communist oppressors"; (2) ex-Batista followers be barred from its armed forces (3 ex-Batista men reportedly had been captured in the Apr. 17 invasion and sentenced to death May 27); (3) all military leaders of any invasion be acceptable to the movement. Prior to the Miami meeting the Ray movement had received a con-

fidential report from its leaders in Cuba that said the Apr. 17 attack had "completely demoralized" Cubans "and especially the militant members of the movement and some of its leaders." The report also said: Many of the movement's top leaders were detained during the mass arrests following the invasion, but the organization remained intact; Cuban government vigilance had made conspiratorial work almost impossible; 90% of Cubans "see intervention" by the U.S. and OAS "as the only solution to the problem"; the movement was "impotent to continue an armed struggle inside" Cuba.

Council Expands. The Revolutionary Council announced in Miami Oct. 12 its expansion to include 14 new members from other anti-Castro groups, some of which were operating underground in Cuba.

The new council members included Higinio Díaz and Francisco Carillo of the Montecristi movement. (Montecristi's president, Dr. Justo Carillo Hernández, had resigned from the council Sept. 30 after complaining of "inactivity" and "sinister interests.")

PW-Tractor Exchange Proposal

Castro offered May 17 to free most of the rebels captured in the Apr. 17 landings in exchange for U.S. bulldozers. His proposal was widely reported as an offer to trade prisoners for tractors. Negotiations then began with a volunteer U.S. committee that sought to raise money for tractors. But the negotiations broke down in June without any exchange.

Castro's Offer. Castro made his initial offer May 17 in a televised address at a rally of peasants in Havana. He said he was willing to exchange 1,214 rebels for 500 U.S. bulldozers. He warned that if his offer were not accepted, the prisoners would "work hard building trenches and fortifications." He qualifiied his offer by adding: "Of course, those [prisoners] . . . who have committed murders cannot be exchanged for anything. . . . The fascist priests we will send for free."

Castro's proposal was accepted May 20 in a telegram sent him by Mrs. Eleanor Roosevelt, United Auto Workers Pres. Walter P. Reuther and Dr. Milton S. Eisenhower, president of Johns Hopkins University. They said they would help raise money as private citizens to pay for the cost of the machinery. Their acceptance of Castro's offer, they said, was made "not as a response to a demand for political ransom but out of common humanity."

A White House spokesman acknowledged May 24 that Pres. Kennedy had taken the initiative in encouraging the formation of the private citizens' committee to raise funds for tractors.

U.S. State Department spokesman Joseph W. Reap said May 20 that despite the "cynicism of the barter proposal" the U.S. would consider issuing licenses to export bulldozers to Cuba if an agreement could be worked out.

PW Committee in U.S. A 10-man committee of captured Cuban rebels arrived in Miami May 20 to negotiate the deal. A Havana radio broadcast said that the prisoners, selected by other captive rebels, had 7 days to reach an agreement. If they failed, they were to return to Cuba.

The prisoner negotiators met in Washington May 22 with Mrs. Roosevelt, Reuther and Eisenhower, who had formally organized their group earlier May 22 as the Tractors for Freedom Committee. Reuther said after the meeting that his committee had given the Cuban delegation "a firm commitment" that the tractors would be sent to Cuba.

The 10 prisoners left Washington and returned to Havana May 27.

The 10 captives: Felix Eloy Pérez, Hugo Luerios, Capt. Luis Moré Delgado, Mirto Collazo Valdes, Waldo Castroverde, Ulises Carbo (son of editor Sergio Carbo of Prensa Libre, a Havana newspaper currently published in Miami), José Perullo, Gustavo García Montes, Ceferino Alvarez Castallón, Reinaldo Pico Ramón.

U.S. Debates Deal. Sen. Wayne Morse asked in a Senate speech May 22 whether the Tractors for Freedom Committee had U.S. government clearance, "and if it does let's openly say that this is part of the foreign policy of the United States" and the committee was acting for the U.S.

Pres. Kennedy May 24 urged Americans to contribute to the Tractors for Freedom Committee fund drive. He said that if the Cuban captives "were our brothers in a totalitarian prison, every American would want to help. I happen to feel deeply that all who fight for freedom—particularly in our hemisphere—are our brothers." Mr. Kennedy said the U.S. had "not been and cannot be a party" to the negotiations involving Castro's offer. But he stressed that the U.S. "should not interfere" with humanitarian efforts of "private citizens [who] seek to help prevent suffering."

Sen. Barry M. Goldwater (R., Arix.), in a Senate speech May 24, criticized Mr. Kennedy's statement as "lending prestige of the [U.S.] government to this surrender to blackmail." The U.S., Goldwater added, would "be lucky if he [Castro] does not wind up demanding 500 hydrogen bombs rather than 500 tractors."

A joint statement condemning the tractors-for-prisoners transaction was issued May 26 by Sen. Everett M. Dirksen (R., Ill.) and Rep. Charles A. Halleck (R., Ind.), minority leaders, respectively, in the U.S. Senate and House of Representatives. The statement said: "The disasterous handling of the Cuban invasion" had resulted in an "enormous loss in [U.S.] prestige abroad. But Pres. Kennedy has compounded that loss by his official sponsorship . . . [of U.S.] blackmail payments to Fidel Castro." Halleck, rejecting the government contention that it was not involved in the fund raising, said that if contributions to the Tractors for Freedom Committee were to be tax-exempt, a "considerable part of the money" would be from the U.S. Treasury.

Sen. Homer E. Capehart (R., Ind.) introduced in the Senate May 25 a resolution expressing opposition to federal tax-exemption for contributions to the committee's campaign. Another resolution introduced by Capehart would, in effect, brand the committee's activities as illegal under the Logan Act, which barred private U.S. citizens from negotiating with foreign governments.

Ex-Vice Pres. Nixon said at an Oklahoma City news conference May 27 that Pres. Kennedy should withdraw his approval of the prisoner-tractor exchange. He called the deal morally wrong and unwise. Nixon said the captives would be released "at the cost of increasing the power of the tyrant of the country." Nixon said Presidential approval of the transaction "would encourage every tinhorn dictator . . . to try to take advantage of America."

Sen Hubert H. Humphrey (D., Minn.), in a Senate speech May 29, backed acceptance of Castro's offer. He said the U.S. "demonstrates strength—not weakness—when it takes action to save human lives." He suggested that the tractor committee's activities "become the springboard from which we launch a tremendous inter-American effort to seek the release of all [Cuban] prisoners. . . ."

U.S. Information Agency Director Edward R. Murrow said at a Washington news conference May 29 that Castro's offer had created "a ground-swell of disgust welling up in Latin America" against the Cuban premier.

(The USIA reported that tractor fund-raising committees had been established in 11 Latin American countries.)

Castro Proposes Alternative. In what appeared to be an alternative offer, Castro demanded May 22 that 1,200 "political prisoners" in the U.S. and elsewhere be exchanged for the Cuban rebels.

Among the prisoners Castro said he wanted returned in his alternative prisoner-for-prisoner proposal were Puerto Rican nationalist leader Pedro Albizu Campos (who was serving a jail term for his part in an unsuccessful Puerto Rican revolt Oct. 30, 1950) and Francisco Molina, a pro-Castro Cuban.° Castro offered to trade the latter for Capt. Manuel Artime, the invasion leader.

Castro also threatened May 22 to call off the tractor-prisoner negotiations if the U.S. continued to refer to the transaction as an exchange for prisoners. Castro said he regarded the tractors as "indemnification" for damage caused by the Cuban invaders.

Castro Raises Demand. Responding to a June 7 Castro offer to negotiate the tractor-prisoners exchange plan in Cuba, 4 U.S. farm machinery experts, representing the Tractors for Freedom Committee, flew to Havana June 13. The committee had said June 8 that it was ready to send the first 100 tractors to Cuba by June 22.

Tractor committee experts: Duane P. Greathouse, 45, a UAW vice president; Roy Bainer, 59, chairman of the University of California's Department of Agricultural Engineering; Prof. John Bruce Liljedahl, 42, of Purdue University's school of agriculture; Asst. Prof. Clarence M. Hensen, 48, of Michigan State University's school of agriculture.

During talks with the U.S. experts June 14, however, Castro demanded

---

*Molina, 29, was sentenced in New York June 29 to a jail term of 20 years to life for his Sept. 21, 1960 accidental fatal shooting of Magdelena Urdaneta, 9, of Venezuela, during a brawl with anti-Castro Cubans in a New York restaurant. Molina also was sentenced to a 5-10 year term for assault with intent to kill an anti-Castro Cuban in the fight and to a 2½-5-year term for 2d degree assault on another. The 3 sentences were to be served concurrently.

1,000 U.S. tractors instead of 500. The experts returned to Washington to report to the committee. The committee June 19 rejected Castro's higher demand and in a cable to Castro, set a June 23 deadline for concluding a deal for 500 tractors.

Wide World
CASTRO June 14

Castro maintained that he had originally asked for 500 bulldozers, worth $28 million. The committee, however, had refused to provide bulldozers lest they be used for military construction. Castro agreed June 14 to drop his demand for bulldozers but insisted that the tractors be equivalent to the value of the bulldozers. He said this would mean 1,000 tractors. (The committee had estimated the cost of 500 tractors at $2½-3½ million.) Castro demanded $28 million in cash or credit if the 1,000 tractors were refused.

In its June 19 cable to Castro, the committee said it had received a list of prisoners to be exchanged but that the total was less than the 1,214 Castro had indicated in his initial offer. Castro had told newsmen in Havana June 13 that 1,173 of 1,197 prisoners captured Apr. 17 were eligible for exchange. The committee said: "This gives greater urgency to our suggestion that the International Red Cross . . . handle . . . the release of the prisoners, including verification of the accuracy of the list."

Castro rejected the 500-tractor limit June 23, the committee's deadline date. Castro said that 500 tractors would not "satisfy even a meager part of the damages suffered because of the cowardly and criminal aggression against Cuba organized by" the U.S. He charged that the committee's offer

was designed "to break off the negotiations and blame" Cuba for it.

Castro June 24 sent the 10 prisoners who had negotiated with the committee in May back to the U.S. to "explain to the North American people the facts." The prisoners conferred in Key West, Fla. June 24, 25 and 27 with John Hooker Jr., executive secretary of the committee (which had dissolved after Castro rejected the June 23 deadline).

Hooker said June 27 that the committee planned no new negotiations because it had disbanded and the captives "cannot negotiate anyway —they are prisoners of a dictator."

Hooker told the prisoners June 30 that the committee would not re-form because "it was impossible to" achieve a fair deal with Castro. Hooker said that by "injecting the concept of indemnity" in the negotiations, Castro "sought to give the impression that the humanitarian intent of private citizens was proof of guilt." Acceptance of Castro's terms, Hooker said, would have been "incompatible with our national honor and with our rôle as private citizens." The committee, Hooker declared, would not give Castro "tractors, money or credit for indemnity—not $28 million or $28."

(2 of the prisoners, Mirto Collazo Valdes and Reinaldo Pico Ramón, both 27, announced in Miami, Fla. July 29 that they had renounced their pledges to return to Cuban prisons. They said they would return to Cuba only to fight Castro. The remaining 8 prisoner-negotiators flew back to Havana July 31. They were accompanied by 4 representatives of a Cuban Families Committee for Liberation who had offered to take the places of Collazo and Pico in prison. The Cuban Families Committee had been formed in Miami July 5 by members of the families of rebels captured in the Apr. 17 landing. The commit-

tee sought money to buy tractors to ransom the prisoners.)

New Exchange Effort. The Cuban Families Committee disclosed in New York Dec. 26 that a representative of the Fiat Co. of Italy was negotiating with officials in Havana to exchange the prisoners for $21 million worth of farm tractors to be delivered by Fiat. The tractors were to be supplied by an international pool of manufacturers. The committee said it would launch a public drive to raise money for the tractors.

Anti-Castro Activity

Drive Vs. Rebels. In early 1961 Castro threw large numbers of troops against rebels in the Escambray and Organos Mountains. Government forces managed to suppress some rebel contingents, but 1961 was marked by a stream of arrests and executions of persons the government charged with sedition and espionage.

An army-militia force of 10,000-15,000 began a drive against Escambray rebels in Las Villas Province in January. The government reported the capture of Fomento Jan. 7 and of Santa Clara Jan. 15 and the defeat of 2 rebel detachments Jan. 16.

A battle between government forces and rebels was reported Feb. 11 between Trinidad and Topes de Collantes in the Escambray Mountains. Both sides were said to have suffered heavy losses. Government troops, numbering about 40,000, reportedly had surrounded the rebels and prevented supplies from reaching them overland. Farmers had been removed from the battle area in a government effort to block rebel food supplies.

Castro said Feb. 11 that "thousands" of weapons supplied to the Escambray rebels by planes from the U.S. had been captured. Government radio commentator José Pardo Llado said Feb. 8 that militia in Las Villas

had captured 86 crates of arms, ammunition and food parachuted to the rebels near Santa Lucia by planes "from Florida." (Pardo, a Castro publicity adviser, fled to Mexico City in March and denounced the Castro régime as dominated by Communists.)

The N.Y. Times reported Apr. 10 that 3,000 militimen had started a new drive against rebels in the Organos Mountains of Piñar del Rio Province. 1,000 of the troops had been shifted from the Escambray Mountains, still considered a rebel stronghold. The Escambray militiamen were reported to have been sent back to Las Villas later to reinforce troops after rebels attacked a militia unit near Sancti Spiritus.

Arrests & Executions. Mass arrests and executions reportedly took place in Cuba prior to, during and after the Apr. 17 landing. Havana newspapers reported Apr. 21 that 14,000 persons had been arrested since Apr. 17. At least 31 persons were said to have been executed Apr. 17-22. (A law approved Jan. 5 authorized death sentences for persons convicted of terrorism and sabotage.)

Pres. Osvaldo Dorticós Torrado Nov. 29 announced a new law providing for immediate death sentences for murderers, invaders and saboteurs. Castro had said Nov. 28 that his régime was introducing "revolutionary justice" against invaders and anti-government plotters by imposing death sentences 48 hours after their capture. Castro singled out for execution saboteurs who burned sugar cane.

Peruvian Pres. Manuel Prado Ugarteche urged the Castro régime Apr. 25 to "suspend the . . . executions."

Costa Rican Foreign Min. Alfredo Vargas announced Sept. 10 that his government had broken diplomatic relations with Cuba in protest against the executions.

The N.Y. Times had reported Jan. 13 that 15,000 persons were political prisoners in Cuba. It said 6,000 "counter-revolutionary suspects" were held on the Isle of Pines and 2,000 in a camp in the Sierra Maestra.

Among arrests, trials and executions reported during 1961:

■ 6 Americans allegedly seeking to join anti-Castro rebels were arrested aboard a yacht in Havana harbor Jan. 7. The government announced the arrests Jan. 19. It said they had sailed from Florida Jan. 6 to join the rebels in Piñar del Rio Province. The Americans, who reportedly had trained at a Florida camp operated by ex-Cuban Sen. Rolando Masferrer, 42, former Batista associate, were identified as Leonard Louis Schmidt, 21, of Chicago, George R. Beck, 24, of Taunton, Mass., Tommy L. Baker, 28, of Dotham, Ala., Donald Joe Green, 28, James B. Beane, 34, of Cedar Falls, N.C. and Eugene Gibson, 32, of Durham, N.C. The men were tried Jan. 28 by a military court in Havana's La Cabana fortress and were sentenced Jan. 31 to 30 years in prison. (Masferrer was arrested at his Miami home Apr. 8 and indicted by a federal grand jury there Apr. 10 on charges of violating the U.S.' Neutrality Act, but the indictment was dropped Nov. 9. He was accused of engineering an unsuccessful 4-boat invasion of Cuba Oct. 4, 1960. Only one boat reached Cuba. Its 28 occupants were eventually captured; 10 of them, including 3 Americans, were executed.)

■ 3 Americans, convicted by a Havana tribunal Dec. 17, 1960, were sentenced Jan. 10 to 10 years in prison on charges of being members of a spy ring that had tapped the telephone wires of Hsinhua, the Communist New China News Agency. The Americans were Edmundo K. Taranske, 30, of New York, Daniel L.

Carswell, 42, of Eastchester, N.Y. and Eustace Dan Brunet, 34, of Baltimore. A 4th defendant, Italian-born Mario Nordio of New York, was deported.

■ A La Cabana military court Jan. 14 sentenced Robert John Gentile, an American, to 30 years in prison after convicting him on charges of plotting against the government. The government press said Gentile and 11 Cuban defendants were members of a "counter-revolutionary organization known as the '28th of January Patriotic Council of Liberation'."

■ William Lex Sante, Julio Casielles Amigo and Orlirio Mendez Perez, ex-Cuban Electric Co. workers who had been captured while trying to flee to the U.S. in a yacht, were executed by a firing squad in La Cabana Jan. 18 after being convicted the same day on charges of bombings that caused a partial black-out of Havana.

■ 3 youths reportedly seeking to assassinate Cuban Bank Pres. Ernesto (Ché) Guevara were captured Feb. 27 after a gun battle with Guevara's bodyguards in front of Guevara's home in the Havana suburb of Miramar. (One of the bodyguards was fatally wounded by one of the youths.)

■ U.S.-born Cuban army Maj. William A. Morgan, 32, and Maj. Jesús Carreras Zayas were executed by firing squads in La Cabana Fortress, Havana Mar. 11 after being convicted of sending arms and ammunition to the Escambray rebels in Sept.-Oct. 1960. Of 12 other person tried for the alleged arms plot, 8 received 30-year jail terms, one, Mario Morin, a prosecution witness, was sentenced to 15 years, and 3 were acquitted.

■ Cuban forces captured an American, Angus McNair, 25, of Miami, and 4 Cubans Mar. 20 after they had landed a boat carrying arms on the north coast of Piñar del Rio Province. 7 Cubans, McNair and another Amer-

ican, Howard F. Anderson, 41, of Yakima, Wash., an owner of Havana service stations, were executed by firing squads in Piñar del Rio Apr. 19.
■ Ex-Agriculture Min. Humberto Sori Marín, Castro's former adjutant general, was executed by a firing squad in La Cabana Apr. 20 on charges of conspiring to assassinate Castro. Sori, a former Havana lawyer, was said to have drafted the Cuban revolutionary law under which military tribunals ordered 600 firing-squad executions. He had been arrested Mar. 13 while transporting arms to Cuba from Cayo Hueso Island. 6 other persons, including an American citizen identified as Rafael Dia Bencom, were executed with Sori.
◪ All but 7 of 14 defendants who had gone on trial in Santiago de Cuba Aug. 21 on charges of participating in "activities against the powers of the state" were sentenced Aug. 24 to prison terms ranging up to 20 years. Those sentenced to 20 years were: student leader Alberto Muller, 23, nephew of Bishop Alfredo Muller; Enrique Gustavo Casuo Perez, alleged leader of anti-Castro rebels in Oriente Province; the Rev. José Luis Rojo Seijas and the Rev. Reinerio Lebroc Martinez, Roman Catholic priests. The remaining defendants received prison terms of from one to 10 years.
■ An American identified as William Patten Tabares, 34, son of William Patten Sr., long-time Cuban resident, was executed by firing squad in Camaguey Sept. 23 after being convicted of activities "against the powers of the state."
■ 5 Cubans were executed by firing squad in La Cabana Sept. 25. They had been convicted with 63 other persons on charges of being recruited by the CIA to sabotage the Castro régime. The 63 received prison terms ranging from 9 to 30 years.

Underground 'Smashed.' The Cuban Department of State Security announced Nov. 7 that the underground anti-Castro Revolutionary Popular Movement had been broken up and most of its 77 leaders arrested. A department communiqué accused the Venezuelan and Italian embassies in Havana of giving refuge to underground leaders and permitting anti-Castro groups to send documents and reports through diplomatic mails.

Catholics Riot. The largest public demonstration yet reported against Castro erupted in Havana Sept. 10 during a Roman Catholic festival at the Church of Charity. One person was killed and 7 injured.

Bishop Eduardo Boza Masvidal had canceled the festival after the government had ruled that the procession, scheduled for the afternoon, had to be conducted before 9 a.m. But about 4,000 worshipers, unaware that the festival had been called off, began assembling inside and outside the church during the day. When they were informed that the procession had been canceled, the worshipers began to march on the Presidential Palace, shouting anti-government slogans. Soldiers and militiamen dispersed the demonstrators by firing over their heads.

An Interior Ministry communiqué issued Sept. 12 accused the Roman Catholic Church in Cuba of plotting to overthrow the régime. It charged that: (1) "Clerical Falangist[s] . . . have received . . . explosives and arms to promote uprisings in Piñar del Rio, Las Villas and Oriente Provinces." (2) Churches and convents were "being used to hide fugitives from revolutionary justice, to hold conspiratorial meetings and even to hide dynamite." (3) Bishop Boza had instigated the Sept. 10 riot, and his Church of Charity was a center of counterrevolutionary activities. (4) Father

Francisco Viera was a main contact between Cuban subversives and the U.S.' CIA.

(Cuban militiamen arrested 176 persons in breaking up an anti-government demonstration during a religious festival near Havana Sept. 17. 18 demonstrators were injured by club-swinging militiamen.

(Premier Castro announced in a TV-radio address Sept. 19 that religious processions would be prohibited and that priests who "conspire against the fatherland" would lose their citizenship.)

Priests Deported. 136 Roman Catholic priests, including Bishop Boza, were arrested by Cuban militiamen in their churches Sept. 15-16 and deported Sept. 17.

The ousted clergymen, who sailed for Spain, included 78 Spaniards, 46 Cubans, 5 Canadians, 4 Dutchmen, one Venezuelan, one Italian and one Hungarian.

They represented nearly half of the clergy remaining in Cuba. About 300 other priests, mostly non-Cubans, who had been jailed following the Apr. 17 invasion, had left Cuba after being released.

Pope John XXIII Sept. 20 protested the deportation of the priests. Addressing pilgrims and tourists at his summer home at Castel Gondolfo, the pope said the church in Cuba was being persecuted.

Defections. Many Cuban government officials and diplomats defected during 1961. Among them was Cuba's first president under the Castro régime, Dr. Manuel Urrutia Lleo. Urrutia and his family were reported Apr. 23 to have taken asylum in Havana in the Venezuelan embassy, which already housed 86 other Cuban refugees. Dr. Juan Orta, ex-head of the Castro régime's defunct Office of the Premier, reportedly sought asylum in the embassy Apr. 13.

The AP reported Apr. 3 that the private secretary to Cuban Pres. Osvaldo Dorticós Torrado, Josefina Lopez Villaverde, had received asylum in Costa Rica's embassy.

Sara Kouri Barreto, 25, sister-in-law of Cuban Foreign Min. Raúl Roa, quit her post Nov. 1 as secretary to Carlos M. Lechuga, Cuba's OAS representative in Washington, and applied for U.S. asylum Nov. 2. Miss Kouri had been Roa's private secretary in Havana before going to Washington.

Among other reported defections:
■ 4 consulate officials in the U.S. resigned and asked for asylum in the U.S. (reported Jan. 9).
■ Cuban Amb.-to-Netherlands Ignacio Fiterre resigned Mar. 15. He said the Castro régime "did not grant justice, freedom and democracy." Fiterre was joined by Miguel Campos, a counselor, and Mrs. Piedad Anduiza, an attaché. 2 members of the Cuban Rotterdam consulate also resigned.
■ Antonio R. Montane, 2d secretary at Cuba's Mexico City embassy, and his wife, Gilda Portela Montane, embassy counsel general, resigned May 11 and sought asylum. They charged that the embassy was a center of Communist activity and a distribution point for Communist propaganda throughout Latin America.
■ Carlos Herrero, Cuban consul general in Montreal, left his post July 18, reappeared in Quebec July 29, and sought asylum.
■ Vitallia Della'torre Perez, Cuban consul general in Buenos Aires, Argentina, and Leonardo Bravo Puig, Cuban consul general in Montevideo, Uruguay, resigned their posts Aug. 4.
■ Jorge Freire Gonzalez, Cuban assistant chargé d'affaires in Manila, sought asylum in the Philippines Sept. 28. He charged that Cuban Chargé d'Affaires Andres Avino Soler was using Cuba's Manila embassy as a center for Communist subversion.

(Philippine Pres. Carlos P. Garcia demanded Soler's recall, and Soler left Manila for Cuba Oct. 13.)

Cuba's Road to Communism

Foes of Castro had charged even before he overthrew the Batista dictatorship in Jan. 1959 that the Castro revolution was a Communist one. These charges, emphatically denied by Castro and most of his supporters, were reinforced by the Castro regime's own actions: mass executions, expropriations, refusal to hold elections, ever-closer identification with the Communist nations, and anti-"imperialist" propaganda on an international scale.

Castro announced in a 1961 May Day talk that he had made Cuba a Socialist country. And he apparently confirmed his enemies' charges by asserting Dec. 2 that he was a "Marxist-Leninist" and was leading Cuba to communism.

Cuba Called 'Satellite.' The U.S. State Department Apr. 3 called Cuba under Castro "a Soviet satellite" and urged it "to sever its links with the international Communist movement." In a 36-page document, written in the White House under the close supervision of Pres. Kennedy, the department called the Castro régime "a danger to the authentic and autonomous revolution of the Americas...."

Among points made in the U.S. document:

■ The Castro revolution, which had overthrown the "Batista tyranny," was first regarded by the Americas "with sympathy." But the "social achievements" of the revolutionary régime had been corrupted by a "consistent betrayal of . . . Castro's pre-revolutionary [political] promises." The régime had become a "repressive dictatorship" marked by "disillusion, persecution, exile and execution" of persons who had supported Castro in his revolution but had "doomed themselves by trying to make his régime live up to his own promises" of freedom, justice and elections.

■ The Communist Party was "the only political party permitted to operate in Cuba" while all other political opposition had been "extinguished." The CP or those "responsive to its influence" "dominated" all aspects of Cuban life. Communists high in political posts included Pres. Dorticós; Cuban CP Secy. Gen. Anibal Escalante, member of Raúl Castro's policymaking group; Dulce Maria Escalona Almeida, director of primary education; Culture Department director Vicentina Atuna.

■ Cuba's "subservience" to "the world Communist bloc" had been marked by (a) "acceptance of the Soviet line on international affairs"; (b) acceptance of "Sino-Soviet arms, equipment, technicians and money"; (c) establishment of diplomatic relations with all Communist nations except East Germany and economic agreements with those countries. Soviet military aid had helped make Cuba's armed force the largest in the Western Hemisphere, with the exception of the U.S.'

■ Cuba had "become a base and staging area for revolutionary activity throughout" the Western Hemisphere. "Cuban influence" had been responsible for recent disturbances in such countries as Venezuela, El Salvador, Nicaragua and Paraguay.

■ The U.S. was determined "to assure future democratic governments in Cuba full . . . support . . . to achieve freedom, democracy and social justice."

Castro Proclaims Socialism. Castro, in a 3½-hour televised address at a May Day gathering in Havana May 1, declared that Cuba was a Socialist country and would no longer hold elections. In the future, Castro said, his régime would depend on direct support expressed by Cubans at rallies.

"Do you need elections?" Castro asked the crowd. "No! No!" his audience shouted. Castro said "the revolution does not contemplate giving the oppressive classes any chance to return to power."

Castro attacked the U.S. for its rôle in the Apr. 17 rebel landing and said: "If Mr. Kennedy does not like socialism, we don't like imperialism, we don't like capitalism."

Castro promised the early enactment of laws barring foreign priests from Cuba and nationalizing private schools, most of which were operated by Roman Catholics.

Earlier, the May Day observance was marked by the burning of an effigy of Pres. Kennedy. There was an all-day parade of marchers carrying anti-U.S. slogans and huge portraits of Karl Marx and of Castro and other Cuban leaders. The parade was reviewed by Castro and foreign delegations including those from the USSR, Red China and Soviet-bloc nations.

(The U.S.' Guantánamo base had its first work stoppage in its 58 years as all but 35 of 3,200 Cuban employes stayed away from their jobs May 1 to participate in the May Day celebrations. They returned to work May 2.)

U.S. State Department spokesman Lincoln White declared May 2 that Castro's speech showed that Cuba "has certainly become a member of the [Communist] bloc." White issued a department statement in which he pointed out that "none of the members of the Communist bloc calls itself Communist. They say they are Socialist and . . . working toward communism." The statement said: Only the USSR and Czechoslovakia "are called Socialist republics"; the others "are called peoples' republics," presumably because "they do not consider themselves to have reached even the Socialist stage yet. It would, therefore, appear that Castro considers Cuba further along the Communist road than some other countries in the bloc."

Cuban Industry Min. Ernesto (Ché) Guevara had anticipated Castro's May Day statement when he (Guevara) declared in a TV-radio address Apr. 30 that Cuba was a "fully socialized country." He emphasized that "the Socialist system is characterized by the possession by the people of the means of production." He described the Cuban revolution as the "first Socialist revolution in America."

Castro told a group of U.S. newsmen who arrived in Cuba June 13 that Cuba had "a Socialist revolution but we did not yet have a Socialist state." He predicted that Cuba would be a Socialist country within 5 years. Castro said he was convinced that a backward country such as Cuba could be developed properly only by socialism and with Soviet-bloc aid. He reported that "several hundred " Soviet-bloc technicians, not including a group of 300 Soviet farm experts who had arrived in Cuba the previous week, were working on the island. The number of Chinese technicians in Cuba was "very small," he declared.

Castro said: The nationwide Committees for the Defense of the Revolution, whose task was to weed out anti-Castro elements, could become the nucleus of a new political party along with the militia and other revolutionary and youth groups; Cuba was transforming agrarian cooperatives into collective or "people's farms" whose workers were government employes; about 15,000-20,000 persons had been arrested during and after the invasion, but "fewer than 2%" were still imprisoned; the number of political prisoners prior to the invasion totaled 5,000; only 5% of the population was actively opposed to the régime.

One-Party Socialism Set. Castro announced July 26 that all Cuban political, military, labor and other revolutionary organizations would be integrated into a single "United Party of Cuba's Socialist Revolution." Speaking in Havana's José Marti Square during celebrations marking the 8th anniversary of his revolutionary 26th of July Movement, Castro said that "this process started months ago." Socialism would be accomplished in Cuba "in a few years," he promised, but he cautioned against trying to rush the process.

Castro promised that small businessmen would not have their businesses taken away by the government. He predicted that small businessmen would discover in "several years" that it was more profitable to work for state enterprises and therefore would quit business voluntarily.

Castro's remarks were made in a 4-hour 7-minute speech at a rally attended by Soviet cosmonaut Yuri A. Gagarin, who assailed the U.S. and promised "the armed help of the Soviet people" in Cuba's struggle for freedom. (See 'Space')

The government Aug. 2 decreed a reorganization of Cuba's labor system. Craft unions were to be replaced by "vertical" or industrial unions, which were to be under the jurisdiction of a single workers' confederation. The announcement said the decree was designed to further a Socialist society.

Castro Plans Communism. Castro announced in a nationwide radio-TV address Dec. 2 that he was leading Cuba to communism. "I am a Marxist-Leninist and will be one until the day I die," Castro declared. The rest of the world "is on the road toward communism" also, he asserted.

Castro said that his new party would lead Cuba through socialism to "a people's democracy, or the dictatorship of the proletariat." Its program

"will be Marxist-Leninist but adapted to conditions" in Cuba. Castro indicated that party members would form a small political élite: although members would be drawn from workers, students, intellectuals, peasants and even the petty bourgeoisie, the party "will be a qualitative and not a quantitative organization." But Castro predicted that "there will be no communism [in Cuba] before 30 years."

Castro, rejecting neutralism, said Cuba must learn from the USSR: "There is no half-way between socialism and imperialism"; "anyone maintaining a 3d position is, in fact, helping imperialism."

In identifying himself as a Marxist-Leninist, Castro explained that he had concealed his Communist beliefs from Cubans and from U.S. friends for years "because otherwise we might have alienated the bourgeoisie and other forces which we knew we would eventually have to fight." Castro said he had become a Communist only after he had risen to power.

Red-Aided 5-Year Plan. Guevara said in a radio-TV address Jan. 6 that Communist nations would assist Cuba in a 5-year "development plan" for solving economic problems resulting from the U.S. decision to halt the import of Cuban sugar. He said the Communist aid was pledged in multilateral agreements signed with Communist-bloc officials during his 1960 Iron Curtain tour. The agreements would enable Cuba to buy goods from any European Communist nation, including the USSR, with the receipts from the 4 million tons of sugar it planned to sell those countries and Communist China in 1961.

Guevara said the plan also would: (1) permit Cuba to send 2,400 persons to Communist nations for technical training; (2) supply Cuba with Polish equipment for a shipyard, for producing shoes and batteries and for

2 slaughter houses; (3) supply Cuba with Chinese Communist equipment for building 24 factories, including plants to manufacture automobile parts, gasoline pumps and paper; (4) extend to Cuba a Czech credit of 40 million pesos for a motor vehicle and motor plant.

Guevara said the USSR had promised to get the formerly U.S.-owned Nicaro and Moa Bay nickel plants back into production soon.

A Bulgarian-Cuban trade agreement signed in Sofia Jan. 12 provided for Bulgaria to give Cuba a $5 million long-term credit and technical aid in building industrial plants by 1965.

Castro announced Jan. 22 that Cuba and the USSR had agreed to exchange 1,000 youths each to work on Cuban and Soviet farms.

Guevara conceded in a radio-TV address Apr. 30 that the government would not be able to accomplish all its economic aims under the current 5-year plan; as a result, some goals would have to be postponed for a 2d 5-year plan. Guevara said some Cuban industrial plants seized from U.S. owners were idle because they lacked raw materials and spare parts formerly received from the U.S. But Guevara said Cuba would: (a) Raise steel production to 500,000 tons annually (The increase was to be effected by USSR construction of a 250,000-ton-capacity steel plant in the Santiago de Cuba area and by modification of the old Antillana de Acero steel factory to produce 250,000 tons by 1963.) (b) Double electric production by the middle 1960s by building 600,000-kw.-capacity generating stations. (Current capacity: 920,000 kw.) (c) Increase annual cement production from one million tons to 2 million. (d) Receive a USSR-built oil refinery (in Santiago de Cuba) with a one million-ton annual production capacity. (e) Import 3,500 textile looms. (Gue-

vara said Cuba hoped to become self-sufficient in the production of textiles.) (f) Produce 2,000 tractors, 5,000 trucks, 3,000 motorcyle motors and 100 stationary motors annually in the plant to be built by Czechoslovakia. (g) Build small-tonnage fishing boats and 10,000-ton ships in the shipyard to be built by Poland.

MiGs Shown. 6 Soviet-built MiG jet fighters were flown over Havana July 25 at the opening of celebrations of the anniversary of the 26th of July Movement. It was the first public display of MiGs by the Cuban air force.

Planes & Ships Hijacked

U.S.-Cuban relations were further strained in 1961 by the "hijacking" of 3 U.S. airliners. 2 were actually flown to Havana but later returned. Cuban planes and ships also were seized by Cubans fleeing Cuba.

Electra Seized in Florida. An Eastern Air Lines Electra, en route from Miami to Tampa, Fla. with 33 passengers and 5 crew members aboard, was captured in the air at gun-point July 24. The gunman, who was one of the passengers, forced the pilot to fly the plane to Havana.

A U.S. request for the return of the plane and those aboard was sent to the Cuban government July 24 through the Swiss embassy.

32 of the plane's passengers and its 5 crew members were released by Cuba and flown to Miami July 25 on a scheduled Pan American Airways DC-6B. The only passenger to remain in Cuba was the man who had seized the Eastern Air Lines plane. He was identified by the FBI as Cuban-born Wilfredo Roman Oquendo, 36, a Miami waiter who had been a U.S. citizen for 18 months.

Premier Castro offered July 26 to return the hijacked plane if the U.S.

promised to return Cuban planes seized and flown to the U.S.

Castro's offer was rejected July 27 by U.S. State Secy. Rusk, who said that claims on seized planes were matters for the courts to decide.

The seized Eastern Air Lines Electra was turned over to Eastern crewmen in Havana Aug. 15 under a U.S.-Cuban agreement. (Negotiations had been conducted through the Swiss embassy in Havana.)

Under the agreement the U.S. returned to Cuba Aug. 15 a Cuban gunboat that had been hijacked July 29 by its 3 crewmen, who brought it to Key West, Fla. and received U.S. asylum. The boat later had been seized under a federal court order granted to the Miami advertising firm of Harris & Co. for non-payment of a debt owed by the Cuban government. The firm relinquished its claim.

Seizure Foiled in Texas. Ex-convict Leon Bearden, 38, of Chandler, Ariz. and his son Cody, 16, seized a $5 million Continental Airlines 707 jet in flight over Arizona en route from Phoenix to El Paso, Tex. early Aug 3. They ordered the pilot, Capt. Byron D. Rickards, 52, to fly to Havana. Rickards insisted that he would have to refuel at El Paso. On landing there, the Beardens freed 61 passengers because a pregnant woman among them appeared hysterical, but 4 volunteers among the passengers and 6 crew members were kept on as hostages.

Authorities on the ground had been alerted by radio before the plane landed, and it was cordoned off. After nearly 4 hours on the ground, the Beardens ordered Rickards to take off. Rickards started to taxi down the field, but the plane was stopped when FBI and border patrolmen shot out its tires and an engine.

2 FBI agents and a border patrolman entered the plane and surprised and subdued the 2 Beardens.

The Beardens were indicted by a federal grand jury in El Paso Aug 7 on charges of kidnaping and transporting a stolen plane across a state line. Bearden was convicted Oct. 18 and sentenced Oct. 31 to life in prison. Cody was sentenced to 5 years in a correctional institution.

DC-8 Hijacked in Mexico. A Pan American World Airways jet DC-8, en route from Mexico City to Guatemala, was seized Aug. 9 by a gunman identified later as Albert Charles Cadon, a Frenchman. Cadon, armed with a pistol, forced the pilot to change course shortly after take-off. The plane, with 81 passengers, including Colombian Foreign Min. Julio César Turbay Ayala, landed in Havana 2½ hours later. The plane was returned to the U.S. the same day.

(Jesus Flores Aguirre, 54, poet and counselor of Mexico's Havana embassy, was shot in his car Aug. 24 in the Havana suburb of Marianao. He died later in a hospital. Mexican authorities, in demanding a "full investigation," asked whether there was a link between Flores' slaying and the Aug. 9 hijacking. Flores had been at the Havana airport when the plane landed.)

Cuban Plane Seizures. A passenger aboard a Cubana Air Lines plane in flight between Varadero Beach and Havana shot and wounded a militia guardsman July 3 and forced the pilot to change course and land in Miami. 13 of the 16 passengers and one of the 4 crewmen reportedly asked for U.S. asylum. Authorities later seized the plane on behalf of Harris & Co.

3 Cubans were shot to death and 6 others were injured when a group of passengers tried to seize a Cuban airliner in flight from Havana to the Isle of Pines Aug. 9. One of 2 militiamen stationed in the plane exchanged shots with the passengers. The pilot,

the 2d militiaman and a passenger were killed. The co-pilot crash landed the plane in Cuba.

Cuban Ships Seized. The 5,085-ton Cuban freighter Bahia de Nipe, which had left Havana Aug. 8 for an undisclosed Communist port with a cargo of sugar, was seized Aug. 14 by 10 defecting crewmen led by the ship's captain, Jorge Augustin Navarro, 46.

The defectors locked up 23 pro-Castro crewmen, radioed the U.S. Coast Guard, sailed the ship into Chesapeake Bay Aug. 17 and anchored off Norfolk, Va. They asked for U.S. asylum.

U.S. District Judge Walter E. Hoffman in Norfolk Aug. 17 authorized the U.S. marshal to seize the freighter after 4 liens were placed against the ship for allegedly unpaid Cuban debts. The Coast Guard, however, refused a U.S. marshal permission to board the vessel Aug. 18 to serve attachment papers. At a court hearing Aug. 19, Hoffman held the Coast Guard action illegal.

State Secy. Rusk Aug. 19 urged the Norfolk court to release the Bahia de Nipe. Rusk said in a letter to Atty. Gen. Robert F. Kennedy that freeing the ship "would avoid further disturbance to our international relations."

Hoffman ordered the ship released Aug. 29; the Bahia de Nipe sailed from Norfolk for Cuba Sept. 15.

39 Cuban passengers, some armed with guns, forced the captain of a 52-foot Cuban motor vessel to reverse course Aug. 29 and bring the boat into Port Antonio, Jamaica. The passengers asked political asylum.

U.S. Hijacking Bill. A bill (S2268) making airplane hijacking a federal offense punishable by imprisonment or death was passed by the U.S. House Aug. 23 and Senate Aug. 28 and was signed by Pres. Kennedy Sept. 5. The bill, which amended the Federal Avi-

ation Act of 1958, also authorized federal penalties for other crimes committed in commercial aircraft.

Cuban-Hemisphere Tensions Mount

Castro to Aid Revolts. Castro asserted in a TV speech Feb. 11 that U.S. encouragement of counter-revolution in Cuba gave him the right to promote revolutions throughout Latin America. In assailing U.S. aid to Cuban refugees, Castro said that if the U.S. had the "right to approve a credit for counter-revolutionary exiles," Cuba had "the right to approve a credit to aid exiled Puerto Ricans and revolutionary exiles in all Latin America."

Castro told a Havana University audience Mar. 14 that Cuba would "form a free Puerto Rican government[-in-exile] as soon as imperialism forms a counter-revolutionary Cuban government-in-exile." He said Cuba would help form exile governments for all Latin American countries that opposed the Cuban revolution.

(The N.Y. Times had reported Jan. 22 that Cuban government officials and leftwing Spanish exiles in Cuba had formed a Communist committee for a Spanish Army of Liberation to spur underground opposition to the governments of Gen. Francisco Franco y Bahamonde in Spain and Premier António de Oliveira Salazar in Portugal. The Times identified the committee's leaders as Capt. Antonio Nuñez Jiminez, Cuban chief of agrarian reform, Santiago Carillo, secretary general of the Spanish Communist Party in exile, and Gen. Alberto Bayo, a Loyalist commander in the Spanish Civil War.)

The U.S. accused Cuba Nov. 1 of trying to "subvert and overthrow the constitutional governments of the Americas." The charge was made by U.S. Amb.-to-OAS deLesseps S. Morrison in a letter sent to OAS Secy.

Gen. José A. Mora and distributed to the OAS' members. The letter contained copies of recent joint communiqués issued by Soviet, Chinese Communist and Cuban leaders. According to Morrison, the communiqués showed that Cuba was "intimately associated with the aims and purposes of the Sino-Soviet bloc."

OAS Acts Vs. Castro. The OAS' Inter-American Defense Board met in Washington Apr. 26 and approved by a 12-1 vote (Cuba opposed, 5 abstentions) a U.S. resolution barring the Cuban representative from attending the board's secret sessions "as long as there exists the present evident alliance of Cuba with the Soviet bloc." Mexico, Brazil, Venezuela, Ecuador and Chile abstained in the vote.

The OAS Council, meeting in Washington Dec. 4, approved by a 14-2 vote a Colombian resolution for the convening of an OAS conference on threats of subversion from Cuba. The resolution, which did not specifically mention Cuba, provided that an OAS foreign ministers conference would open Jan. 10, 1962 to "consider the threats to peace and the political independence of the American states that may emerge from an intervention of extra-continental powers." (The negative votes were cast by Cuba and Mexico. 5 nations abstained: Argentina, Bolivia, Brazil, Chile and Ecuador.)

The Cuban government Dec. 18 rejected a request of the OAS' Inter-American Peace Commission to visit Cuba to investigate Peruvian charges of Communist subversion against other Latin American countries and of human rights violations. In a note rejecting the request, Cuban Amb.-to-OAS Carlos M. Lechuga assailed the commission's "abject submission and repugnant immorality." The commission rejected Lechuga's note as "not received" on the ground that it con-

tained "injurious concepts of the OAS, the Peace Commission" and its member states.

13 States End Ties. Antagonism toward the Castro régime gained momentum in the Americas during 1961. Opposition to Castro, which was most marked among government and business leaders, seemed to be making inroads into widespread admiration for Castro and support for his revolution that persisted in Latin America.

These 13 Western Hemisphere countries had ended ties with Cuba by Dec. 14: Colombia, Costa Rica, the Dominican Republic, El Salvador, Guatemala, Haiti, Honduras, Nicaragua, Paraguay, Peru, Venezuela, Panama and the U.S.

The Colombian and Panamanian action was taken after Castro had charged Dec. 8 that the 2 countries' régimes were "government[s] of traitors" and accomplices of Yankee imperialism and had invited them to end diplomatic relations with Cuba. Colombian Pres. Alberto Lleras Camargo, announcing Colombia's break Dec. 9, charged that Castro had provoked it to create an atmosphere of danger to coincide with U.S. Pres. Kennedy's scheduled Dec. 17 visit to Colombia. Panama's cabinet voted the break Dec. 14 but postdated it to Dec. 8, the day of Castro's attack.

El Salvador's break with Cuba had been reported Feb. 28. Honduras' cabinet decided on the same action Apr. 24 after a pro-Castro crowd of 2,000 demonstrated before the presidential palace in Tegucigalpa and denounced Honduran Pres. Ramon Villeda Morales as a traitor for deciding on the break. Venezuela ended relations with Cuba Nov. 12, but Venezuelan Pres. Rómulo Betancourt said in a Nov. 12 radio-TV address that Venezuela would not permit its territory to be used "as a base for any army . . . to invade" Cuba.

Rapprochement Urged. The UN General Assembly Apr. 22 approved by a 59-13 vote (24 abstentions) a 7-nation Latin American resolution calling on "all member states to take . . . peaceful action . . . to remove existing tensions" between the U.S. and Cuba. Prior to the vote the Assembly had deleted from the resolution a U.S.-supported paragraph urging OAS member states to "lend their assistance" in ending the dispute.

The U.S. Apr. 28 rejected an offer announced by Cuban Pres. Dorticós Apr. 27 that the U.S. and Cuba negotiate their differences in order to establish "peaceful coexistence, diplomatic relations and even friendly relations, if the [U.S.] so desires." The announcement, signed by Castro and Dorticós, was read by Dorticós to a group of foreign diplomats summoned to the presidential palace in Havana. Announcing the U.S. rejection, U.S. State Department spokesman Lincoln H. White said: "The basic problem in Cuba is communism in this hemisphere"; "communism in this hemisphere is not negotiable" because it "is a multilateral problem, not a bilateral problem."

Adolf A. Berle Jr., chairman of a task force to coordinate U.S. policies on Latin America, had said Mar. 2, after talks with Brazilian Pres. Jânio Quadros in Brasília, that a Cuban proposal to have Latin American countries mediate the U.S.-Cuban dispute was not "feasible" because the U.S. regarded the matter as of concern to the whole Western Hemisphere and not a problem that related only to the 2 governments. (Quadros said May 10: He opposed "any foreign intervention, direct or indirect," in Cuba; a Cuban government that resulted from "a clearly manifest interference by a foreign power" would never be recognized by Brazil; Brazil would "join in measures to preserve the integrity of the continent short of intervention in any country of the hemisphere.")

DR Rôle Probed. The Cuban government called on the UN Security Council in New York and the OAS in Washington Nov. 21 to consider Cuban charges of U.S. "aggression" in the Dominican Republic.

In a letter to Security Council Pres. Valerian A. Zorin of the USSR, Cuba's UN delegate, Mario Garcia-Inchaustegui, charged that the U.S. had planned "armed intervention" in the DR to uphold the last vestiges of the Trujillo dictatorship. Zorin agreed to the emergency meeting, despite British and French requests for time to study Cuba's plea.

The UN Security Council opened debate on the Cuban charges Nov. 22, took no action and ended the debate Nov. 28.

The OAS Council held a special meeting Nov. 22 on the Cuban charges. Cuban delegate Carlos M. Lechuga said the U.S. naval action threatened to throw the Americas back to the days of constant military interventions. U.S. delegate deLesseps S. Morrison replied that the "friendly presence" of the U.S. warships was designed to avert "political disintegration." DR delegate Virgilia Diaz Ordonez called Lechuga's charge an intervention in Dominican affairs.

Alliance for Progress

A U.S. initiative for economic and social betterment in Latin America was pledged by John F. Kennedy Jan. 20 in his inaugural address:

"To our sister republics south of the border, we offer a special pledge—to convert our good words into deeds —in a new Alliance for Progress—to assist free men and free governments in casting off the chains of poverty."

The Alliance for Progress plan was approved overwhelmingly by the countries of Latin America at the Inter-American Economic & Social Conference in Punta del Este, Uruguay in August.

Stevenson's Tour. U.S. Amb.-to-UN Adlai E. Stevenson visited 10 South American countries in June as Pres. Kennedy's personal representative. The main purpose of the trip was to discuss the Alliance for Progress with Latin American leaders.

Stevenson returned to Washington June 22 and reported to Mr. Kennedy June 23. At a news conference afterwards, Stevenson said:

■ Economic and political conditions had worsened in Latin America since his 1960 visit to the area. "Popular discontent" had increased. Government "stability" had declined. Communist and rightwing subversion had "clearly increased."

■ The countries he visited displayed "a more universal awareness of the necessity for making social reforms." The Alliance for Progress "attracted universal enthusiasm and approval."

■ "The principle of non-intervention is a religion in Latin-America." Insofar as the Apr. 17 Cuban invasion "can be considered American intervention, it is unpopular in South America."

■ Latin Americans favored the "objectives of the revolution in Cuba" but were largely unaware that the revolution had been "betrayed" and preverted by communism.

In his report to State Secy. Dean Rusk (made public July 24), Stevenson said Latin America must meet the "threat from Cuba" with "acts on the home front to eliminate the attraction of Castro-communism for ordinary people." Stevenson said: He had told Latin American officials that the U.S. "has less to lose or suffer . . . from Sovietization of Cuba than Latin America; that the immediate potential victims are the structurally vulnerable Latin American democracies, and that they are already the objects of indirect Communist aggression"; but reliance solely on "political or economic action" against the Cuban government "might well increase rather than eliminate the Communist threat by investing Fidelismo with an aura of martyrdom." "Communism and its alter-ego, Castroism, thrive on urban slum conditions and rural insecurity"; the poor, "who are no longer passive politically, must see reasons for hope in life under free institutions; when they do, statesmanship of the Americas will be more free to plan for the longterm development of their countries without being distracted by fears of rebellion." Aid and strong argument would be needed to reverse Latin feeling that the "Cuban peasant has acquired his place in the sun."

Kennedy's Pledge. Pres. Kennedy Aug. 5 promised Latin America more than $1 billion in U.S. economic aid during the first year of the Alliance for Progress. The pledge was made in a message read to the opening session of the Inter-American Economic & Social Conference at Punta del Este, Uruguay.

Guevara Offers Castro Aid. In an address to the conference Aug. 8, the head of the Cuban delegation, Industry Min. Ernesto (Ché) Guevara, declared that the Castro régime would support the Alliance for Progress "if it truly contributed to growth of the economies [of Latin America], to increasing national production and not to the extraction of more natural resources from Latin America, if it furthers social justice, not another imperialist war."

Guevara said Cuba would promise not to "export revolution" to other American countries if Cuba received a pledge of "non-intervention to go ahead with our work." The conference,

Guevara contended, was a "political meeting conceived against Cuba and what it represents as an example for all America."

Guevara said Cuban security forces had thwarted a U.S. plot to assassinate Maj. Raúl Castro, Cuba's armed forces minister, during ceremonies in Santiago de Cuba July 26. The alleged plot, according to Guevara, called for the firing of mortar shells into the U.S.' Guantánamo naval base by Cuban-based anti-Castro rebels as a pretext for U.S. armed intervention in Cuba.

Asst. State Secy. Robert F. Woodward denied Guevara's accusations.

Alliance Charter Adopted. The Alliance for Progress Charter, implementing the U.S.' Alliance for Progress plan, and an accompanying Declaration of Punta del Este, summarizing the charter's goals, were adopted by the conference Aug. 16. The documents were signed by representatives of the U.S. and 19 other American countries at the conclusion of the conference Aug. 17. Cuba refused to sign the documents.

After signing for the U.S., Treasury Secy. Dillon said the U.S. had "no intention of making any of its funds available" to Cuba as long as it was "under the control of a foreign power —namely, the Soviet Union."

Other Developments

Economic Reforms. A reorganization of government agencies to provide more efficient control of Cuba's economy was decreed Feb. 24. The reform created a Ministry of Industry headed by National Bank Pres. Ernesto (Ché) Guevara. The ministry was put in charge of all industry. It was to carry out a 1962-66 industrialization program announced by Guevara Feb. 22. The ministry replaced the Cuban Petroleum and Cuban

Mining Institutes, the Department of General Administration of Sugar Mills and other industrial departments formerly under the National Agrarian Reform Institute's direction. Land and agriculture control remained under the jurisdiction of the agrarian institute, headed by Capt. Antonio Nuñez Jimenez.

The decree abolished the Bank of Foreign Trade and replaced it with a Ministry of Foreign Trade headed by Maj. Alberto Mora Becerra, president of the disbanded institution. The new ministry was authorized to negotiate commercial pacts and control foreign exchange operations formerly handled by the National Bank. The ministry also assumed the functions of collecting customs duties and taxes.

The National Bank of Cuba was reorganized and was empowered to conduct the country's entire domestic and foreign banking operations. Ex-Commerce Min. Raúl Cepero Bonilla was named president of the bank.

An Interior Commerce Ministry was established to handle "distribution of products, control of prices and management of all [nationalized] commercial firms."

Economic Shortages. Premier Fidel Castro said at the opening session of a 3-day "criticism" meeting of the National Production Congress in Havana Aug. 26 that Cuba was beset by shortages of food and consumer products. He blamed the deficiencies on increased buying power rather than on a drop in production.

The meetings, attended by officials connected with farm and industrial production, were ordered by the government and the National Directorate of the ORI (Integrated Revolutionary Organization), Castro's new political party formed by a merger of his 26th of July Movement and the Communist Party. Industry Min. Ernesto Guevara reported at the meeting Aug. 27 that many Cuban factories had been forced to shut down because of shortages of raw materials and spare parts previously obtained from the U.S. Only 2 of 40 state industrial enterprises had met their 1961 production goals, but some of them had exceeded their 1960 production figures, Guevara said.

Castro had announced in a speech July 4 the rationing of animal and vegetable fats because of shortages. 12 carloads of lard bound for Cuba had been seized in West Palm Beach, Fla. June 28 under a court order obtained by a Miami advertising company to help pay a judgment against the Cuban government's tourist agency. Castro said that Cuban fat consumption, normally 40¼ pounds per person a year, would have to be restricted to one pound of oil and one pound of lard per person a month. Castro said the USSR had made 10,000 tons of lard available to Cuba but that this was not enough.

UN Aids Farm Study. The UN Special Fund's governing council May 24 approved an agricultural study for Cuba despite U.S. objections. The fund was to pay $1,157,600 and Cuba $1,878,000 for a study of Cuban livestock production, soil classification, conservation and crop diversification. The U.S.' fund delegate, Philip M. Klutznick, reportedly had objected that the program probably would fail because (a) Cuba had slaughtered 2½ million of a "basic herd" of 6 million cattle that was to have been used in the livestock program and (b) the Santiago de Las Vegas planning station, site of the project, had fewer trained agronomists than Cuba had reported when applying for the aid (the missing agronomists reportedly had fled Cuba).

Currency Reform. Cubans exchanged old pesos for newly-printed ones Aug. 6-7 under a government order issued Aug. 5. The replacement was at par. (The Cuban peso, officially valued at one U.S. dollar, sold for 20¢-25¢ on the free market.) The government said the measure was aimed at removing inflationary pressure of "hundreds of millions of pesos" held outside Cuba.

The new currency law allowed each family to exchange up to 200 old pesos for 200 new ones. Old peso holdings above that amount were to be turned in and held by the government in special accounts that could not be withdrawn in exchange for the new money until one week later. Pesos unexchanged after Aug. 7 were proclaimed worthless.

The government had barred all ships and planes from entering the country from Aug. 5 to 7 to prevent old pesos from coming in pending completion of the exchange. All stores and businesses were closed, except those engaged in essential services. Civilans employed at the U.S.' Guantánamo base were barred from going to their jobs until Aug. 8.

Castro said in a TV address Aug. 8 that most persons would be limited to no more than 10,000 pesos. Any amount turned in above 10,000 would be confiscated, he said, because it would prove violation of an anti-hoarding law decreed in May. Old persons and cripples would be exempt from the 10,000-peso ceiling and would be permitted up to 20,000.

The Cuban National Bank reported Aug. 18 that the Aug. 6-7 money exchange had reduced currency circulation in Cuba to about 447,900,000 new pesos, or 40% of the 1.187 billion old pesos that had been circulating before the exchange. More than 497 million pesos (42% of the pre-exchange total) had been permanently cancelled, according to the bank.

(The government Mar. 15 had ordered Cubans and foreign residents to surrender all their foreign currency and foreign securities to the Cuban National Bank by Apr. 3. The currency and securities were exchanged for Cuban pesos.)

Education Nationalized. Cuba's ernment would provide free educaunder a decree approved by the cabinet June 7. The law said the government would provide free education and "take over all educational establishments." The decree affected about 1,000 private schools, some of which, including Roman Catholic schools, already had been seized by the militia.

The opening of a government drive to teach peasants to read and write had started May 15 as the first of 100,000 volunteer teachers left Havana for the Cuban interior.

Secret Police Ministry. A Ministry of Interior was established June 7. The ministry was headed by Maj. Ramiro Valdes, ex-secret police agency chief, and organized along the lines of the USSR's central security agency. It supervised a secret police network that included the Technical Police and the uniformed National Revolutionary Police. The ministry also was in charge of the Committees of the Defense of the Revolution.

U.S. Airlift Blocked. The U.S. State Department announced July 21 that Pres. Kennedy had approved a plan by which the U.S. government would pay about $350,000 from emergency foreign aid funds to airlift more than 20,000 Cubans seeking to flee to the U.S. The plan was blocked July 22 by a Cuban Transport Ministry order forbidding Pan American World Airways to increase the number of its Miami-Havana flights to more than 2 daily.

Emigration Curbed. Restrictions on emigration were announced Sept. 14 and resulted in the immediate cancellation of more than 15,000 plane tickets held by U.S.-bound Cubans. Under the new regulations, prospective emigrants were required to obtain new police exit permits valid for only one week.

1962

One of the most threatening Soviet-U.S. confrontations of the cold war took place late in 1962 in the Caribbean. The U.S., announcing that it had discovered aggressive Soviet missile bases under construction in Cuba, blockaded the island and demanded the removal of the bases. For 6 days the world lived in fear of an armed clash between the 2 great powers. Then the Soviet government announced that it would dismantle the launching sites and withdraw its missiles. By the year's end, U.S. reconnaissance planes had verified Russia's removal of 42 medium-range missiles and 42 jet bombers.

Pres. Kennedy Oct. 23 signs proclamation for blockade to bar Cuban import of offensive arms   (Wide World photo)

U.S. Demands Soviets Dismantle Rocket Sites

An atmosphere of crisis was generated throughout the U.S. Oct. 21-22 by reports that Pres. John F. Kennedy and other Administration leaders had interrupted their campaigning for the Nov. 6 Congressional elections to return to Washington for a series of meetings held under the tightest military and political secrecy.

The events in Washington took place during a time of growing friction between the U.S. and Cuba's revolutionary Communist government. The tension had been heightened by repeated reports of the presence of Soviet troops, jet aircraft and missiles on the island.

It was announced in Washington Oct. 22 that the President would address the nation later that day on a matter of the highest national importance. An estimated 50 million Americans were at their radios and TV sets as he began to speak, informing them that the USSR had begun to build offensive missile sites in Cuba.

JFK Acts Vs. Threat. Pres. Kennedy informed the American people Oct. 22 that the USSR had begun building offensive missile launching bases in Cuba. He announced that he had ordered Cuba blockaded to prevent further deliveries of offensive weapons to the island. He called on Soviet Premier Nikita S. Khrushchev to dismantle the rocket sites, to withdraw Russian missiles from the island and to end this 'provocative threat to world peace."

The President disclosed the Cuban missile threat and the U.S.' response to it in a nationwide radio-TV address begun at 7 p.m. EDT. Mr. Kennedy said: "This government . . . has maintained the closest surveillance of the Soviet military buildup on the island of Cuba. Within the past week unmistakable evidence has established the fact that a series of offensive missile sites is now in preparation on that . . . island. The purpose of these bases can be none other than to provide a nuclear strike capability against the Western Hemisphere."

Mr. Kennedy said that the U.S.' first "hard" information on the Cuban bases, placed in his hands Oct. 16, showed that 2 types of installation were under construction. "Several of them include medium-range ballistic missiles capable of carrying a nuclear warhead for a distance of more than 1,000 nautical miles," he declared. "Additional sites not yet completed appear to be designed for intermediate-range ballistic missiles capable of traveling more than twice as far and thus capable of striking most of the major cities in the Western Hemisphere. . . ." "In addition," he said, "jet bombers, capable of carrying nuclear weapons, are now being uncrated and assembled in Cuba while the necessary air bases are prepared."

Mr. Kennedy warned that this "urgent transformation of Cuba into an important strategic base by the presence of these large long-range and clearly offensive weapons . . . constitutes an explicit threat to the peace and security of all the Americas."

The President expressed contempt for "the repeated assurances of Soviet spokesmen . . . that the arms buildup in Cuba would retain its original defensive character and that the Soviet Union had no need or desire to station strategic missiles on the territory of any other nation." He quoted from a Sept. 11 Soviet government statement in which the USSR had defended its Cuban arms deliveries and said: "That statement was false." He cited a similar avowal made to him personally by Soviet Foreign Min. Andrei A. Gromyko during a White House meeting, Oct. 18—that Soviet assistance to Cuba " 'pursued solely the purpose of contributing to the defense capabilities of Cuba' "—and commented: "That statement also was false."

Mr. Kennedy declared that the Soviet-Cuban action had upset the current nuclear balance between East and West. He said: "For many years both the Soviet Union and the United

States . . . have deployed strategic nuclear weapons with great care, never upsetting the precarious *status quo* which insured that these weapons would not be used in the absence of some vital challenge." "But this secret, swift, extraordinary buildup of Communist missiles in an area known to have a special and historical relationship to the United States and the . . . Western Hemisphere . . . —this sudden, clandestine decision to station strategic weapons for the first time outside Soviet soil—is a deliberately provocative and unjustified change in the *status quo* which cannot be accepted by this country if our courage and our commitments are ever to be trusted again, by either friend or foe."

The President made it clear that "our unswerving objective . . . must be to prevent the use of these missiles against this or any other country; and to secure their withdrawal or elimination from the Western Hemisphere." He said he therefore had ordered the following "initial steps" be taken:

"First, to halt this offensive buildup, a strict quarantine on all offensive military equipment under shipment to Cuba is being initiated. All ships of any kind bound for Cuba from whatever nation or port, will, if they are found to contain cargoes of offensive weapons, be turned back. This quarantine will be extended if needed to other types of cargo and carriers.

"We are not at this time, however, denying the necessities of life as the Soviets attempted to do in their Berlin blockade of 1948.

"2d, I have directed the continued and increased close surveillance of Cuba and its military build-up. . . ."

"3d, it shall be the policy of this nation to regard any nuclear missile launched from Cuba against any nation in the Western Hemisphere as an attack by the Soviet Union on the United States requiring a full retaliatory response upon the Soviet Union.

"4th, as a necessary military precaution, I have reinforced our base at Guantánamo, evacuated today the dependents of our personnel there and ordered additional military units to be on a stand-by alert basis.

"5th, we are calling tonight for an immediate meeting of the Organ of Consultation under the Organization of American States to consider this threat to hemisphere security and to invoke Articles 6 and 8 of the Rio Treaty in support of all necessary action. . . .

"6th, we are asking tonight that an emergency meeting of the [UN] Security Council be convoked without delay to take action against this latest Soviet threat to world peace.

"Our resolution will call for the prompt dismantling and withdrawal of all offensive weapons in Cuba under the supervision of UN observers before the quarantine can be lifted.

"7th, and finally, I call upon Chrmn. Khrushchev to halt and eliminate this clandestine, reckless and provocative threat to world peace and to stable relations between our 2 nations.

"I call upon him further to abandon this course of world domination and to join in an historic effort to end the perilous arms race and to transform the history of man."

Mr. Kennedy concluded with the following words to his U.S. audience: "My fellow citizens, let no one doubt that this is a difficult and dangerous effort on which we have set out. No one can foresee precisely what course it will take, or what costs or casualties will be incurred. . . . The cost of freedom is always high, but Americans have always paid it."

USSR Rejects Blockade. The Soviet government rejected Pres. Kennedy's blockade warning and declared Oct. 23 that any aggressive American actions toward Cuba or its sea lanes to the Soviet bloc could result in thermo-

nuclear war. The USSR's initial official reaction came in a government statement made public Oct. 23. Addressing itself to Pres. Kennedy's radio-TV speech, the statement said:

"The peoples of all countries must be clearly aware that, undertaking such a gamble [by blockading Cuba, the U.S.] . . . is taking a step along the road of unleashing a thermonuclear world war. Cynically flouting international standards, . . . the United States usurped the right . . . to attack ships of other states on the high seas— *i.e.*, to engage in piracy.... The Soviet government reaffirms that all weapons of the Soviet Union serve and will serve the purposes of defense against aggressors. . . . The Soviet Union will continue to discharge this mission with all firmness and consistence. . . . [The U.S.] accuses Cuba of allegedly creating a threat to the security of the United States. But who will believe that Cuba could create [such] a threat . . .? It is hypocrisy, to say the least, to allege that small Cuba can encroach on the security of the United States. . . .

"As to the Soviet Union's assistance to Cuba, it is aimed solely at enhancing Cuba's defense potential. As stated on Sept. 3 [in a joint Soviet-Cuban communiqué] . . . these arms and military equipment were designed exclusively for defensive purposes. . . . [The U.S.] demands that military equipment Cuba needs for self-defense should be removed from Cuban territory, a demand which . . . no state which values its independence can meet."

"Peace-loving states cannot but protest against the [U.S.] piratic actions . . . with regard to the ships sailing toward Cuba's shores. . . . The establishment of an actual blockade of the Cuban shores by the United States is a provocative move. . . . The United States . . . arrogates the right to demand that states report to it how they organize their defense, [and] what do they carry in their ships in the open sea. The Soviet government resolutely rejects such claims. . . ."

(The Soviet Defense Ministry announced later Oct. 23 that it had been ordered to cancel all military leaves and to defer the discharges of senior age groups from the strategic rocket forces, the anti-submarine warfare branch and the submarine fleet. It also was announced in Moscow Oct. 23 that Marshal Andrei A. Grechko, commander-in-chief of Warsaw Treaty forces, had conferred with senior staff members of the other Soviet-bloc nations and had issued orders to "raise the military preparedness of the troops and fleets.")

Castro Assails Blockade. Cuban Premier Fidel Castro assailed the U.S. quarantine Oct. 23 as a total blockade that was "a violation against the sovereign rights of our country and all the peoples." Addressing the Cuban people on TV, Castro said U.S. reasons for the blockade were "absolutely without foundation." The move, he contended, had been designed to conceal a surprise U.S. attack against Cuba, but Cuba's armed forces were on a "war footing" to meet it.

Insisting that Cuban arms were "defensive," Castro declared: "We will acquire the arms we feel like acquiring and we don't have to give an account to the imperialists. Cuba has the right to arm itself, and we have had to do so. What would have occurred if we had not been armed at the time of Girón Beach [the 1961 invasion]?" As for the U.S.' demands that Cuba submit to inspection to assure removal of the threat to its security, Castro warned that the inspectors "had better come ready for combat."

Cuba's armed forces had been alerted Oct. 22 immediately following Pres. Kennedy's TV address. A com-

muniqué issued Oct. 23 said: "Day dawned with the nation on a war footing ready to repel aggression"; "the heroic defenders of the revolution are at their posts ready to die in defense of their homeland."

Latins Back U.S. The Council of the Organization of American States Oct. 23 approved a 4-point U.S.-sponsored resolution authorizing "the use of force" in carrying out the U.S. arms quarantine against Cuba. The resolution was adopted by a vote of 19-0 at an emergency meeting held in Washington. Uruguay abstained because its delegation had not received instructions from its government at the time of voting, but Uruguay approved the resolution Oct. 24 and made the OAS action unanimous.

The OAS Council earlier Oct. 23 had granted itself emergency powers on the ministerial level as an Organ of Consultation under terms of the 1947 Inter-American Treaty of Reciprocal Assistance, which provided "for effective reciprocal assistance to meet armed attacks against any American state and in order to deal with threats of aggression from them."

The OAS resolution said there was "incontrovertible evidence" that Cuba had "secretly endangered the peace of the continent by permitting the deployment of intermediate and middle-range missiles capable of carrying nuclear warheads on its territory by the Sino-Soviet powers."

The resolution: (1) Urged "the immediate dismantling and withdrawal from Cuba of all missiles and other weapons with any offensive capability"; (2) recommended that OAS states "take all measures, individually and collectively, including the use of armed force . . . to ensure that . . . Cuba cannot continue to receive from the Sino-Soviet powers military material . . . which may threaten the peace and security of the continent and to prevent the missiles in Cuba with offensive capability from ever becoming an active threat to the [continent's] peace and security"; (3) announced the OAS Council's decision "to inform" the UN Security Council of this resolution in "the hope that the Security Council will . . . dispatch UN observers to Cuba at the earliest moment"; (4) announced the OAS Council's plan "to continue to act provisionally as the Organ of Consultation and to urge member states to keep the Organ of Consultation duly informed of the measures adopted in accordance with Paragraph 2 of this resolution."

Several Latin American nations offered the U.S. support and assistance at the height of the crisis:

▶ Haitian Pres. François Duvalier, in a communiqué issued Oct. 25 and made public Oct. 26, put his country's harbor and airfield facilities at the disposal of U.S. naval and air units involved in the quarantine operation.

▶ 2 Argentine warships sailed from Puerto Belgrano Oct. 28 for the Caribbean to join U.S. naval units in enforcing the quarantine. The country's armed forces were reported Oct. 25 to have been alerted for action "in any theatre of operation."

▶ Mexican Pres. Adolfo López Mateos, in a letter sent to Cuban Pres. Osvaldo Dorticós Torrado Oct. 23, said Mexico "cannot view with indifference a Latin-American country establishing facilities for the utilization of the most destructive weapons of all times."

▶ Guatemalan Pres. Manuel Ydígoras Fuentes, in supporting Pres. Kennedy's actions on Cuba, said Oct. 23 that the Guatemalan army would serve "any place where its obligations call it."

(In contrast to other Latin leaders, Brazilian Premier Hermes Lima upheld "Cuba's right to carry out its political experiment." Speaking Oct. 23 at a pro-Cuban demonstration outside the Foreign Ministry building in Rio de Janeiro, Lima said: "Latin America is not bound in serfdom to any international interests that oblige us to maintain a type of régime in which the people do not find conditions for their material improvement and national liberation.")

UN Meets on Charges. The UN Security Council convened in New York Oct. 23 to hear conflicting charges of aggression brought before it by the U.S., Cuba and the USSR.

The 11-member Council was called into session by Soviet Amb.-to-UN Valerian A. Zorin, Council president for October, on the basis of requests received Oct. 22 from U.S. Amb.-to-UN Adlai E. Stevenson and Oct. 23 from Cuban Amb. Mario Garcia-Inchaustegui.

Stevenson's request asked that the Council deal with "the dangerous threat to the peace and security of the world caused by the [USSR's] secret establishment in Cuba . . . of launching bases and the installation of long-range ballistic missiles capable of carrying thermonuclear warheads to most of North and South America."

The Cuban request asked the Council "to consider the act of war unilaterally committed by . . . [the U.S.] in ordering the naval blockade of Cuba."

Stevenson was the first of the 3 contending representatives to address the Council. Tracing the history of the U.S.' postwar efforts to seek a *modus vivendi* with the USSR and of the Soviet Union's frustration of these efforts, Stevenson declared that the USSR had sought to carry "the cold war into the heart of the Americas." Its success, he said, had brought the transformation of Castro's government into a "puppet" régime and the emplacement of Soviet missiles in Cuba.

Stevenson declared that the USSR had turned Cuba into a Communist "bridgehead and staging area," into "a base for putting all of the Americas under the nuclear gun." Stevenson warned the Council: "Since the end of the 2d World War, there has been no threat . . . so profound. . . . The action we take may determine the future of civilization."°

Garcia-Inchaustegui, who spoke immediately after Stevenson, derided the U.S.' alleged classification of missiles and bases into "good and bad" depending on their country of origin. But he devoted his address primarily to a denunciation of the quarantine: "The American blockade . . . is an act of war. It is the use of force by a great power against the independence of our homes. It is a criminal act violating the [UN] Charter and the principles of our organization. . . . The reply of our people and of our government to the imminent armed attack of the United States has been general mobilization."

Zorin, following Garcia-Inchaustegui, reiterated the Soviet warning that "in embarking upon an open adventure of this kind, . . . [the U.S. was] making a step towards the unleashing of a world thermonuclear war." Zorin rejected Stevenson's version of postwar history and of recent events in Cuba. He declared: "The Soviet delegation . . . officially confirms the statement already made . . . in which

it was said that the Soviet government has not directed and is not directing to Cuba any offensive armaments."†

U.S. Blockades Cuba

Quarantine Imposed. The U.S. blockade of Cuba was imposed beginning at 2 p.m. Greenwich time (10 a.m. EDT) Oct. 24 under a proclamation signed by Pres. Kennedy Oct. 23. In the proclamation, Mr. Kennedy ordered "the forces under my command . . . to interdict . . . the delivery of offensive weapons and associated material to Cuba."

The proclamation listed "prohibited material" as "surface-to-surface missiles; bomber aircraft; bombs; air-to-surface rockets and guided missiles; warheads for any of the above weapons; mechanical or electronic equipment to support or operate the above items; and any other classes of material hereafter designated" by the Defense Secretary.

The Defense Secretary was ordered "to prevent the delivery of prohibited material to Cuba, employing the [U.S.'] land, sea and air forces . . . in cooperation with any forces that may be made available by other American states." He was authorized to designate "prohibited or restricted zones" and "prescribed routes" "within a reasonable distance of Cuba" and to "issue such [other] directives as he deems necessary. . . ."

The proclamation provided that: "Any vessel or craft which may be proceeding toward Cuba may be intercepted and may be directed to identify itself, its cargo, equipment and stores . . . to submit to visit and search, or to proceed as directed. Any vessel or craft which fails or refuses to respond to or comply with directions shall be subjected to being taken

---

*A U.S. draft resolution submitted to the Council by Stevenson called for (1) "the immediate dismantling and withdrawal from Cuba of all missiles and other offensive weapons"; (2) the sending to Cuba of a UN "observer corps to assure and report on compliance with this resolution"; (3) "termination of the measures of quarantine . . . [on UN] certification of compliance"; (4 U.S.-Soviet talks "on measures to remove the existing threat to the security of Western Hemisphere and the peace of the world."

†A Soviet draft resolution presented by Zorin called for (1) condemnation of the U.S. action toward Cuba; (2) elimination of the blockade; (3) ending of U.S. "interference into the domestic affairs" of Cuba; (4) the U.S. and USSR "to establish contacts and enter into negotiations for the purpose of normalizing the situation and thereby removing the threat of war."

in custody. Any vessel or craft which it is believed is *en route* to Cuba and may be carrying prohibited material or may itself constitute such material shall . . . be directed to proceed to another destination of its own choice and shall be taken in custody if it fails or refuses to obey. . . . Force shall not be used except in case of failure or refusal to comply with directions, . . . regulations or directives . . . after reasonable efforts have been made to communicate them to the vessel or craft, or in case of self-defense. . . . Force shall be used only to the extent necessary."

U.S. Caribbean Buildup. A U.S. military build-up had been taking place in the Caribbean area since at least early October. Military strength at the Guantánamo Bay Naval Base in Cuba, previously numbering 3,300 naval personnel and several hundred Marines, was doubled, and dependents of servicemen at the base were sent to the mainland Oct. 23.

It was reported Oct. 23 that 40 ships and 20,000 men (including 6,-000 Marines) already in the Caribbean for what were described as annual naval training maneuvers off Puerto Rico had been diverted to blockade duty.

An additional Navy squadron of 12 F-4B Phantom 2 jet fighters had been transferred to the Key West, Fla. area early in October in what was admitted to be an answer to the Soviet shipment of MiG jet fighters to Cuba. The arrival of several thousand combat-equipped Marines at the Key West Naval Base was also reported. Planes of the 19th Air Force arrived at MacDill Air Force Base, Tampa, Fla. Oct. 17, and the base was put on "SAC alert."

The Defense Department disclosed Oct. 23 that U.S. military forces throughout the world had been put "on alert." Discharges of Navy and Marine personnel were halted by Defense Secy. Robert S. McNamara Oct. 23 under an executive order issued that day. The Strategic Air Command announced Oct. 24 that all leaves had been canceled and all personnel on leave recalled to duty. 24 Air Force Reserve troop carrier squadrons, 8 troop carrier wing headquarters and 6 aerial port squadrons, totaling 14,214 men, were recalled to active service Oct. 28 for up to 12 months.

Tightened security regulations kept all but a minimum of news about the military response to the Cuban crisis from reaching the public.

Blockade Clash Averted. No direct confrontation of U.S. and Soviet naval power took place off Cuba despite the U.S. blockade.

Defense Secy. McNamara had reported Oct. 23 that as many as 25 Russian and Soviet-bloc vessels were *en route* to Cuba and would be subject to halt and search by the U.S.' naval blockade. Pentagon spokesmen reported Oct. 24, a few hours after the blockade took effect, that 12 of the Soviet ships had changed course but that the others were still steaming toward Cuba. The Pentagon officials expressed the view that the diverted vessels presumably had been carrying weapons or military supplies banned by the blockade proclamation.

The first contact between a Soviet vessel and the U.S. blockaders took place off Cuba early Oct. 25. It was announced officially by the Defense Department only after Rep. James E. Van Zandt (R., Pa.) had reported the contact to newsmen after attending a State Department briefing in New York on the Cuban situation. The Defense Department reported that U.S. naval units had halted the Soviet tanker *Bucharest* but had permitted it to continue toward Havana without search after "the Navy satisfied itself that no prohibited material was aboard this particlular ship."°

The 2d contact of the blockade was made Oct. 26 when U.S. vessels, among them the destroyer *Joseph F. Kennedy*, named after the President's late brother, intercepted and boarded the Lebanese-registered freighter *Marucla*, bound from Riga, USSR to Havana under Soviet charter. *The Marucla's* officers cooperated with the U.S. boarding party, and the vessel was permitted to continue to Havana after its cargo was checked. The *Marucla*, according to later reports, was a U.S.-built World War II Liberty ship owned by Greek shipping interests through a Panamanian company.

Diplomatic Maneuvering

Khrushchev for Summit. In his first public statement since the start of the crisis, Khrushchev Oct. 24 indorsed a proposal that he and Pres. Kennedy meet to seek a resolution of the crisis.

His statement, made in reply to an appeal for such a meeting by Bertrand Russell, British philosopher and pacifist, came at a time when worldwide fears had been expressed that the crisis might result in war.†

Khrushchev, in a letter written to Russell and made public in Soviet broadcasts and press dispatches, said: "I understand your worry and anxiety" over U.S. "aggressive actions" "in pushing the world to the brink of war" by blockading Cuba; "I . . . assure you that the Soviet government will not take any reckless decisions, will not let itself be provoked by the unwarranted [U.S.] actions . . . and will do everything to eliminate the situation fraught with the irreparable consequences which has arisen in connection with the aggressive [U.S.] action"; if the U.S. continued its "piratic actions" the USSR "shall have to resort to means of defense against an aggressor to defend our rights and in-

ternational rights"; "the question of war is so vital that we should consider useful a top-level meeting in order to discuss all the problems which have arisen, to do everything to remove the danger of the unleashing of a thermonuclear war."

Thant Plea for Peace. Acting UN Secy. Gen. U Thant appealed to the U.S., Soviet and Cuban governments Oct. 24 to suspend all potentially aggressive actions and begin negotiations for a peaceful resoluton of the crisis.

Thant's plea was made at the request of more than 40 non-aligned nations whose delegations had been caucusing at UN headquarters. It was delivered in personal messages addressed by Thant to Pres. Kennedy and Khrushchev.

Appearing before the Security Council later that day to report on his action, Thant said that his messages to Mr. Kennedy and Khrushchev had asked that "in the interest of international peace and security all . . . refrain from any action which may aggravate the situation and bring with it the risk of war." "This involves

---

*The Defense Department said it was obvious that the Bucharest was carrying petroleum, which was not a "prohibited material" under Pres. Kennedy's blockade order. A Soviet tanker arrived in Havana Oct. 26 and was welcomed by a dockside rally honoring its crew for "running" the blockade. There was some confusion over the fact that Cuban spokesmen referred to the tanker as the Vinnitsa and photographs of the vessel showed this name on its side in Cyrillic letters.

†Demonstrations protesting the U.S. blockade took place in New York, Washington, London, and La Paz, Montevideo and Rio de Janeiro Oct. 23-28. The American demonstrations were led by pacifist groups, among them Women Strike for Peace, the Student Peace Union and the Fellowship for Reconciliation, that contended that the U.S. should have sought UN redress before proclaiming the Cuban quarantine. The British demonstrations were staged by the Committee for 100, an anti-nuclear test group; 150 of its members were arrested when they defied government prohibitions to stage a "Hands Off Cuba" rally in Trafalgar Square Oct. 27. The Latin American demonstrations were largely the work of pro-Castro organizations. In Moscow, demonstrating school children paraded outside the U.S. embassy Oct. 24 and splashed the building with ink; large numbers of organized Communist protests against the blockade took place throughout the Soviet bloc.

U.S. Defense Department released above illustration Oct. 28 and identified it as a low-level photo taken Oct. 23 of medium-range ballistic missile site under construction in San Cristobal area of Cuba.
(Wide World)

on the one hand the voluntary suspension of all arms shipments to Cuba, and also the voluntary suspension of the quarantine measures involving the searching of ships bound for Cuba." "I believe," Thant wrote, "that such voluntary suspension for a period of 2 to 3 weeks will greatly ease the situation and give time to the parties concerned to meet and discuss with a view to finding a peaceful solution of the problem."

Thant took advantage of his appearance before the Council to address a separate appeal to Castro and Pres. Osvaldo Dorticós Torrado of Cuba to help the UN's effort to find "a way . . . out of the present impasse." Thant said: "I believe it would also contribute greatly . . . if the construction and development of major military facilities and installations in Cuba could be suspended during the period of negotiations."

Replies to Thant. The first formal U.S. and Soviet acknowledgement that both sides desired a peaceful res-

olution of the crisis came in the replies by Khrushchev and Pres. Kennedy Oct. 25 to U Thant's appeal.

Conciliatory replies were delivered to Thant from both the U.S. and Soviet leaders. The responses differed, however, in that Khrushchev fully accepted the Thant proposal whereas Mr. Kennedy accepted the suggestion for preliminary talks without mentioning Thant's request for suspension of the blockade. The messages were made public when they were read in the Security Council by Stevenson and Zorin.

The Khrushchev message said: ". . . I understand your concern about the situation obtaining in the Caribbean since the Soviet government also considers this situation as highly dangerous and requiring an immediate interference by the United Nations. I am informing you that I agree with your proposal, which meets the interests of peace."

The Kennedy message: ". . . [The] threat was created by the secret introduction of offensive weapons into Cuba, and the answer lies in the removal of such weapons. . . . You have made certain suggestions and have invited preliminary talks. . . . Amb. Stevenson is ready to discuss promptly these arrangements with you."

U.S. Says Bases Speeded. White House spokesmen said Oct. 25 that the U.S. could not relax its blockade or surveillance of Cuba as long as reports showed, as they had, that "work is continuing on these [missile] bases."

The White House officials who refused to be identified, drew attention to a paragraph in Pres. Kennedy's Oct. 22 speech that said that "further action will be justified" if work on the bases was not halted. They added: "It is self-evident that the quarantine [blockade] will continue. There are still Soviet ships headed toward Cuba, and the only way this government can get precise information on some of those ships or the cargo they are carrying is through the quarantine. . . ."

A White House statement Oct. 26 declared: "The development of ballistic missile sites in Cuba continues at a rapid pace. . . . The activity at these sites apparently is directed at achieving a full operational capability as soon as possible." It charged that "there is evidence that as of yesterday, Oct. 25, considerable construction activity was being engaged in at the intermediate range ballistic missile sites. . . . Since . . . Oct. 23, missile-related activities have continued at the medium-range ballistic missile sites. . . ." The Oct. 26 report concluded: "In summary, there is no evidence to date indicating that there is any intention to dismantle . . . these sites. On the contrary, the Soviets are rapidly continuing their construction of missile support and launch facilities and serious attempts are un-

der way to camouflage their efforts."

State Department officials again drew newsmen's attention Oct. 26 to Pres. Kennedy's threat of "further action" against Cuba if "these offensive military preparations continue." And U.S. Amb.-to-UN Stevenson said Oct. 27 at a New York meeting of 13 representatives of NATO, CENTO and Rio Treaty nations that the U.S. was ready to take military action against the Cuban bases "in a brief space of time" unless they were dismantled quickly.

Stevenson-Zorin Exchange. U.S. Amb.-to-UN Stevenson Oct. 25 displayed to the UN Security Council aerial photos purporting to show offensive missile bases in Cuba. Stevenson produced the pictures after Soviet delegate Zorin had refused a direct reply to Stevenson's question: "Do you . . . deny that the USSR has placed and is placing medium- and intermediate-range missiles and sites in Cuba, Yes or no—don't wait for the translation—yes or no?"

Zorin, chairman of the Council for its 3d session on the Cuban crisis, replied: "I am not in an American courtroom, . . . and therefore I do not wish to answer a question which is put to me in the fashion in which a prosecutor puts questions. In due course, sir, you will have your reply."

Stevenson asserted that he was "prepared to wait for my answer until Hell freezes over, if that is your decision. I am also prepared to present evidence in this room." He thereupon had the photos brought in.°

Zorin called the photos "forged" and refused to look at them. He re-

*Aerial photos of Cuban rocket and aircraft bases had been shown by Defense Secy. McNamara to newsmen at a briefing Oct. 22, but the Defense Department refused to allow them to be published. Apparently due to a misunderstanding, a U.S. embassy official in London distributed some of the photos Oct. 23, and the Pentagon therefore released 14 of the photos Oct. 24. Objects marked "launch pads," "missile transporter-erectors," "missile trailers," "probable erector-launcher equipment" and even "missiles on launchers" were visible on the photos.

minded the Council that Stevenson had shown photos dealing with the abortive 1961 invasion of Cuba and that these later had been branded as false by the U.S. press.

Stevenson, upholding the photos' authenticity, challenged Zorin to ask Cuba to permit UN observers to inspect the missiles shown in the pictures. Prior to his exchange with Zorin, Stevenson had declared that the U.S. had been forced to take "prompt action" against Cuba to counteract the "speed" and "stealth" with which the Soviet Union had built missile sites in Cuba.

Zorin challenged the U.S.' claim of proof that the USSR had built missile bases in Cuba. The U.S. had "no such facts in its hands except this falsified information" in the CIA photos, Zorin declared.

The Council meeting was adjourned without setting a new date, presumably to permit U Thant to continue his efforts to solve the crisis.

Pledges Vs. Clash. Khrushchev and Pres. Kennedy pledged Oct. 26 to take all steps to avoid a direct naval confrontation on the U.S.' blockade line around Cuba. Their pledges were made in responses to a 2d appeal from U Thant.

Thant's new appeal was cabled to the U.S. and Soviet leaders Oct. 25. It expressed his "grave concern that Soviet ships already on their way to Cuba might challenge the quarantine . . . and produce a confrontation at sea between Soviet ships and United States vessels, which could lead to an aggravation of the problem" and "destroy any possibility of the discussions I have suggested." Thant asked Khrushchev "to instruct the Soviet ships already on their way to Cuba to stay away from the interception area for a limited time only, in order to permit discussions. . . ." He urged Mr. Kennedy to instruct

U.S. vessels in the Caribbean "to do everything possible to avoid direct confrontation with Soviet ships in the next few days to minimize the risk of any untoward incident."

The Khrushchev and Kennedy replies were transmitted to Thant Oct. 26 and made public later that day. The 2 leaders accepted his appeal for caution in the Caribbean, but each strongly denounced the other for the continuation of aggressive policies in the Cuban situation.

Khrushchev declared in reply that "if any conflict should arise on the approaches to Cuba—and this may become unavoidable as a result of the piratical measures taken by the United States—this would beyond question seriously complicate the endeavors . . . to put an end, on the basis of negotiation, to the critical situation that has now been thrust on the world by the aggressive actions of the United States. We therefore accept your proposal and have ordered the . . . Soviet vessels bound for Cuba but not yet within the area of the American warships' piratical activities to stay [temporarily] out of the interception area."

Mr. Kennedy said that "if the Soviet government accepts and abides by your request . . . you may be assured that . . . [the U.S.] will accept and abide by your request that our vessels in the Caribbean . . . 'avoid direct confrontation with Soviet ships'."

Cuba-Turkey Deal Urged. Khrushchev offered Oct. 27 to withdraw all Soviet missiles and offensive weapons from Cuba if Pres. Kennedy would pledge the U.S. to a similar withdrawal from its bases in Turkey.

Khrushchev's proposal, made in a letter transmitted to Mr. Kennedy and broadcast by Moscow radio, contained the admission that the USSR had sent to Cuba missiles and other weapons deemed offensive by the U.S.

The Soviet leader said he understood Mr. Kennedy's concern for U.S. security as a result of the Cuban situation. He added: "You want to make your country safe. This is understandable, but Cuba, too, wants the same thing. All countries want to make themselves safe. But how are we, the Soviet Union, . . . to assess your actions which are expressed in the fact that you have surrounded with military bases the Soviet Union; surrounded with military bases our allies; . . . have stationed your rocket armament there. . . . You are worried by Cuba . . . because it is a distance of 90 miles by sea from the coast of America. But Turkey is next to us. Our sentries walk up and down and look at each other. Do you . . . have the right to demand security for your own country and . . . not acknowledge the same right for us?"

Khrushchev continued: "I therefore make this proposal: We agree to remove those weapons from Cuba which you regard as offensive weapons. We agree to do this and to state in the United Nations this commitment. Your representatives will make a statement to the effect that the United States . . . will evacuate its analogous weapons from Turkey. . . . And, after that, persons entrusted by the United Nations Security Council may check on the spot the fulfillment of the pledges made."

Khrushchev made it clear that the USSR had not relinquished control to Cuba of the weapons sent to the island. He said: "The weapons on Cuba . . . which you say alarm you are in the hands of Soviet officers. Therefore any whatsoever accidental use of them to the detriment of the United States is excluded."

U.S. Rebuffs Proposal. A White House statement issued Oct. 27 in reply to the broadcast version of Khrushchev's Cuba-Turkey offer rejected any such *quid pro quo* as a basis for resolving the crisis.

The statement, while it did not reject the specific terms mentioned by Khrushchev, declared that "the first imperative must be to deal with this immediate threat, under which no sensible negotiations can proceed." Asserting that "work on these offensive weapons is still proceeding at a rapid pace," the statement said that it was "the Western Hemisphere countries and they alone that are subject to the threat that has produced the current crisis." "As to the proposals concerning the security of nations outside this hemisphere," the U.S. declared that efforts to secure "properly inspected arms limitation" could be resumed "as soon as the present Soviet-created threat is ended."

(Turkish officials Oct. 28 expressed gratification at the U.S.' rejection of Khrushchev's Oct. 27 offer to remove Soviet missiles from Cuba in exchange for the withdrawal of U.S. missile bases in Turkey. Foreign Min. Feridan Cemal Erkin said: "Missiles in Turkey are entirely defensive. Turkey cannot be the subject of bargaining in this matter.")[*]

JFK Accepts Earlier Offer. Pres. Kennedy disclosed later Oct. 27 that a prior Khrushchev offer to settle the Cuban crisis would be accepted by the U.S. unless it had been superceded by the proposal for a Cuban-Turkey "deal."

The President's disclosure was made in a message in which he replied to a hitherto-unreported letter that had been sent to him by Khrushchev Oct. 26, before the Cuba-Turkey proposal. In this earlier letter, Mr. Kennedy said, Khrushchev had offered to re-

---

[*]There were 15 Jupiter IRBM launching sites in Turkey. It had been reported repeatedly that the U.S. was studying withdrawal of the Jupiters, which were liquid-fueled and considered obsolete and vulnerable to attack.

move the Soviet missiles from Cuba under UN supervision if the U.S. suspended its blockade and gave assurances that it would not invade the island.

Mr. Kennedy's reply to Khrushchev said, in part: "As I read your letter, the key elements of your proposals—which seem generally acceptable as I understand them—are as follows:

"(1) You would agree to remove these weapons systems from Cuba under appropriate United Nations observation and supervision; and undertake, with suitable safeguards, to halt the further introduction of such weapons systems into Cuba.

"(2) We, on our part, would agree —upon the establishment of adequate arrangements through the United Nations, to insure the carrying out and continuation of these commitments— (a) to remove promptly the quarantine measures now in effect and (b) to give assurances against an invasion of Cuba. I am confident that other nations of the Western Hemisphere would be prepared to do likewise."

The President reiterated that "the first thing that needs to be done . . . is for work to cease on offensive missile bases in Cuba and for all weapons systems in Cuba capable of offensive use to be rendered inoperable, under effective United Nations arrangements. Assuming this is done promptly, I have given my [UN] representatives . . . instructions that will permit them to work out . . . an arrangement for a permanent solution to the Cuban problem."

Khrushchev Accepts Withdrawal. Khrushchev informed the U.S. Oct. 28 that he had ordered that Soviet missiles be withdrawn from Cuba and that all offensive Soviet bases on the island be dismantled under UN inspection.

Khrushchev's decision was announced in a letter in which he re-

plied to Pres. Kennedy and assured him that the USSR stood by its offer for a settlement. Khrushchev's letter, broadcast by Moscow radio and distributed by the Tass news agency Oct. 28 before its delivery, said:

"I regard with great understanding your concern and the concern of the peoples of the United States of America in connection with the fact that the weapons you describe as offensive are formidable weapons, indeed. . . .

"In order to eliminate as rapidly as possible the conflict which endangers the cause of peace . . . the Soviet government, in addition to earlier instructions on the discontinuation of further work on weapons construction sites, has given a new order to dismantle the weapons, which you describe as offensive, and to return them to the Soviet Union.

". . . The Soviet government decided to render assistance to Cuba with the means of defense against aggression, only with means for the purposes of defense. We have supplied the defense means which you describe as offensive means. We have supplied them to prevent an attack on Cuba....

"I . . . trust the statement you made in your message on October 27th, 1962, that there would be no attack, no invasion of Cuba, and not only on the part of the United States, but also on the part of other nations of the Western Hemisphere, as you have said in the same message of yours. Then the motives which induced us to render assistance of such a kind to Cuba disappear.

"As I had informed you in the letter of October 27, we are prepared to reach agreement to enable [UN] representatives . . . to verify the dismantling of these means. . . ."

Pres. Kennedy declared Oct. 28 that Khrushchev's agreement to withdraw Soviet missiles from Cuba under UN verification was a "statesmanlike decision" and an "important and constructive contribution to peace." Mr. Kennedy made known his acceptance of Khrushchev's Oct. 28 note even before the message had been translated officially and delivered to the White House. In a preliminary statement issued by the White House, the President said he would continue to maintain contact with UN headquarters and U Thant on the elaboration of "reciprocal measures to assure peace in the Caribbean area."

In a more formal note transmitted to Khrushchev and made public later Oct. 28, the President said he had replied "at once" because of "the great importance I attach to moving forward promptly to the solution of the Cuban crisis." "I consider my letter to you of Oct. 27 and your reply today as firm undertakings . . . which should be promptly carried out. . . ."

Castro Demands Guantánamo. Cuban Premier Castro asserted in a broadcast Oct. 28 that unless the U.S. withdrew from its naval base at Guantánamo, "the guarantees of which Pres. Kennedy speaks against invasion of Cuba will not exist."

Castro also demanded: (a) "the end of the economic blockade and all measures of commercial pressure" exercised against Cuba by the U.S. "in all parts of the world"; (b) "the end of all subversive activities, dropping and landing of arms and explosives by air and sea, organization of mercenary invasions, infiltration of spies and saboteurs"; (c) "the end of [U.S.-based] pirate attacks against Cuba"; (d) "the end of all violations of air and naval space by North American military aircraft and ships."

Quarantine Suspended. The U.S. Defense Department announced Oct. 29 that the naval "quarantine" of Cuba would be suspended for 48 hours beginning at daylight Oct. 30. The announcement said that the blockade would be "voluntarily sus-

pended" by the U.S. in response to a request by U Thant, who had announced Oct. 28 his acceptance of an invitation from Castro to go to Havana to discuss a resolution of the crisis. Spokesmen reported that no Soviet vessels were known to be near the blockade line, although a few vessels of other registry were in the area with cargoes of wheat from Soviet ports. It was made clear that the blockading vessels would remain in position on the edge of the zone they had patrolled since Oct. 24.

The suspension of the U.S.' aerial surveillance of Cuba was announced in Washington Oct. 30. Administration officials said that the program of photo reconnaissance flights over Cuba's missile bases and airfields had been suspended for at least 48 hours. The suspension was ordered despite what these officials termed "no conclusive evidence" that the Soviet missile bases were being dismantled.

There were, however, several indications Oct. 30 that the Administration had evidence of Soviet action to destroy the bases. Asst. Defense Secy. Arthur Sylvester told newsmen that although he had no definite report on the results of the latest reconnaissance flights, it was "fair to assume" that no evidence existed of the USSR's failure to carry out its pledge.

Thant-Castro Talks. U Thant conferred in Havana with Premier Castro, Pres. Dorticós, Foreign Min. Raúl Roa and other Cuban leaders Oct. 30-31 in an effort to work out arrangements for UN supervision of the dismantling and removal of Soviet missile bases in Cuba. His trip was without definite outcome.

Thant's Havana trip had been proposed by Castro Oct. 27 in a message that requested "direct discussions on the present crisis, prompted by our common purpose of freeing mankind from the dangers of war." Castro's note had said Cuba would be willing to suspend construction and develop- ment of the missile bases during nego- tiations if the U.S. "desists from threats and aggressions against Cuba, including the naval blockade. . . ."

Thant returned to New York Oct. 31 and reported that, although he and Castro had held "fruitful negotations," he had obtained no agreement in his attempt to get Cuba to permit UN ob- servers to check on the removal of the bases. Thant said, however, that he had been "reliably informed that the dismantling of the missiles and their installations was already in progress and that the process should be com- pleted by Friday [Nov. 2]." Thant said arrangements were "in hand" for the equipment's "shipment and re- turn" to the USSR.

U.S. Renews Blockade. The U.S. naval blockade and aerial recon- naissance of Cuba were resumed at dawn Nov. 1. The resumption of the blockade and surveillance was ordered by the White House after U Thant's return from Havana.

Defense Department spokesmen re- ported Nov. 1 that U.S. planes had returned "without incident" from the first of their new surveillance flights over Cuba although the Cuban gov- ernment had threatened to shoot down any U.S. aircraft entering Cuban air- space on a surveillance mission.

(Pentagon spokesmen were quoted Nov. 1 as saying they believed that a U-2 reconnaissance jet shot down over Cuba Oct. 27 had been hit by a Soviet SAM-2 ground-to-air missile, similar in type to the U.S.' Nike-Her- cules. They were of the opinion that the missile, if used, had been fired by Soviet technicians. The body of Maj. Rudolf Anderson Jr., 35, of Del Rio, Tex., pilot of the downed U-2, was returned to the U.S. and was buried Nov. 6 in Greenville, S.C. Gen. Thomas S. Power, SAC commander,

attended the funeral and told news- men that Anderson and Maj. Richard S. Heyser had piloted the reconnais- sance planes that had obtained "the first conclusive" evidence of the Soviet missile buildup in Cuba.)

Castro Bars Inspection. Premier Castro Nov. 1 rejected the USSR's proposals for international inspection on Cuban soil of the withdrawal of Soviet missiles.

Castro, in a radio-TV address, spe- cifically turned down Soviet proposals for (a) UN on-site inspection of dis- mantling of the bases, (b) UN inspec- tion of Soviet ships bound for Cuba and (c) International Red Cross Com- mittee inspection of Soviet ships in Cuban ports. Castro also rejected a plan for UN aerial inspection of Cuban missile sites as suggested by U Thant during his talks in Havana.

Reading what he described as the entire transcript of his Oct. 30 con- ference with Thant, Castro said he had told Thant: "We do not accept inspection demands"; the Soviet promise to withdraw the missiles "should be enough"; "the Soviet de- cision was serious, and the arms are being withdrawn"; "what right has the United States to ask us to submit to inspection?"; Cuba was opposed to the U.S.' pledge of "no invasion" in exchange for establishment of inspec- tion machinery because the U.S. "has no right to invade Cuba and we can- not negotiate on the basis of a promise that a crime will not be committed."

Castro, concluding his speech, said: The Soviet Union "had the right to take away the weapons that were theirs," and Cubans "respect the So- viet decision"; "if we have misunder- standings with the Soviets, it must be discussed only among the principals"; with these misunderstandings it is important to remember what the So- viet has done for us, . . . and we thank him [Premier Khrushchev] for it"; the USSR gave Cuba arms "and did not charge us for it"; "the missiles were not Cuba's, . . . [therefore] we have no reason to be discontented."

USSR Destroys Missile Bases

JFK Reports Dismantling. Pres. Kennedy reported to the nation Nov. 2 that the USSR had begun dismantling the Cuban missile bases.

Speaking in a radio-TV broadcast from the White House, Mr. Kennedy also disclosed that the U.S. would accept Red Cross inspection of Soviet vessels bound for Cuba but that aerial surveillance of the island would be maintained until acceptable means had been found to verify the complete withdrawal of the missiles and destruction of their launching sites.

Mr. Kennedy said, in part:

"I want to . . . report on the conclusions which this government has reached on the basis of yesterday's aerial photographs . . . namely, that the Soviet missile bases in Cuba are being dismantled, their missiles and related equipment are being crated and the fixed installations at these sites are being destroyed. The United States intends to follow closely the completion of this work through a variety of means, including aerial surveillance, until . . . an equally satisfactory international means of verification is effected. While the quarantine remains in effect, we are hopeful that adequate procedures can be developed for international inspection of Cuba-bound cargoes. The International Committee of the Red Cross . . . would be an appropriate agent in this matter. The continuation of these measures in the air and sea until the threat to peace posed by these offensive weapons is gone is in keeping with our pledge to secure their withdrawal or elimination from this hemisphere. . . . Progress is now being made toward the restoration of peace in the Caribbean. . . ."

(The aerial photos cited by the President were made public by the Defense Department Nov. 3. Taken by low-flying planes, the photos showed the removal of missile erectors, trailers and launch-preparation facilities at 4 sites at which mobile medium-range ballistic missiles had been discovered. The photos were of facilities located near Sagua la Grande, 130 miles east of Havana, and San Cristóbal, 50 miles southwest of Havana. Pentagon spokesmen said that other photos showing demolition work at intermedate-range ballistic missile bases had not been released because they had been taken from different angles and would not be meaningful to the public.)

Inspection Agreements. U.S. and Soviet negotiators reached agreement on 2 forms of verification of the elimination of the Cuban missile threat.

These agreements, produced by intensive U.S.-Soviet negotiations carried out at UN headquarters in New York, provided for (1) Red Cross inspection of Soviet vessels *en route* to Cuba, and (2) U.S. observation at sea of crated Russian missiles being shipped back from Cuban ports to the USSR.

The UN negotiations had been convened by U Thant on his return from Havana Oct. 31.

The agreement for Red Cross inspection of Cuba-bound Soviet vessels was reported to have been reached Nov. 2. It was said to have been advanced by the USSR after the rejection by Castro of the earlier Soviet proposal for the use of Red Cross-designated observers in mobile inspection teams, for on-site verification of the base dismantling and for dockside checks of the weapons being returned to the USSR.

The 2d agreement, for U.S. observation at sea of the Soviet missiles being removed from Cuba, was reached at the UN Nov. 7 and made public by the Defense Department later that day. The department's announcement said: "The Soviet Union has reported ships are leaving Cuba with missiles aboard. Arrangements are being made with Soviet representatives for contact with these ships by United States naval vessels and for counting the missiles being shipped out."

Chief participants in the UN negotiations were Thant, U.S. Amb.-to-UN Stevenson, John J. McCloy, named by Pres. Kennedy Oct. 29 to be chairman of a special coordinating committee on Cuban policy, Carlos M. Lechuga, who replaced Garcia-Inchaustegui Nov. 1 as Cuban delegate to the UN, Soviet Amb.-to-UN Zorin, Deputy Foreign Min. Vasily V. Kuznetsov, who had arrived in the U.S. Oct. 28 to act as Khrushchev's special envoy on the Cuban problem, and First Deputy Premier Anastas I. Mikoyan.

(Mikoyan had landed in New York Nov. 1 *en route* to Havana for talks with Castro. Meeting newsmen at the Soviet mission to the UN, he declared that "if every side will adhere to their documents [on the missile withdrawal], this problem will be solved.")

Mikoyan-Castro Talks. Mikoyan flew to Havana Nov. 2 after his stopover in New York to begin talks with Castro.

Soviet sources declared officially that the purpose of Mikoyan's visit was to discuss trade and economic questions. But it was assumed widely that Mikoyan had been sent by Khrushchev to persuade—or force— Castro to comply with the inspection pledges given by Khrushchev in his exchange of messages Oct. 28 with Pres. Kennedy.

No communiqué was issued after the talks, which took place Nov. 2-5. Havana radio, reporting on the Nov. 3 conference, said merely that Castro and Mikoyan had discussed "international tensions."

Mikoyan declared in an interview with the Havana Communist newspaper *Hoy* Nov. 3 that these "are difficult days for the Cuban people . . . [but] the Soviet people are with Cuba body and soul." Mikoyan, denouncing "aggressive acts of imperialists against the Cuban people," said he "wish[ed] to be one more soldier of revolutionary Cuba."[°]

Castro, commenting on Mikoyan's visit, had said in a Nov. 1 speech: "We will sit down with the Soviet Union as equals and discuss in the light of reason and principle on a government-to-government basis because, above all, we are Marxist-Leninist."

Khrushchev: Missiles 'Out.' The removal of Soviet missiles from Cuba was announced by Khrushchev Nov. 7 at a Kremlin reception in observance of the 45th anniversary of the Bolshevik Revolution. He did not specifically state that the missiles all had left Cuba, but this was believed to have been the intent of his remarks.

In an impromptu toast to peace at the Kremlin gathering, Khrushchev asserted that the USSR had emplaced 40 missiles in Cuba but that "we have taken our rockets out and they probably are on the way."

Khrushchev added that although the Cuban crisis was not yet ended, it no longer threatened the world with war. He declared that at the beginning of the crisis "we were very close —very, very close—to a thermonuclear war." He asserted that "if there had not been reason then we would not be here tonight and there might not

---

*Mikoyan continued his talks with Castro despite the death of his wife, Mrs. Ashken Lazaryevna Mikoyan, 64, in Moscow Nov. 3. Mikoyan's son, Sergei, 27, left Havana Nov. 4 to return to Russia to attend the funeral.

have been elections in the United States."

(Reports that U.S. reconnaissance planes had photographed the loading of Soviet missiles in Cuban ports were confirmed inadvertently Nov. 6 by Amb.-to-UN Stevenson as he emerged from negotiations with Soviet envoys at the UN. Stevenson, asked by newsmen about the progress of his talks, replied that "the missiles are moving out, the talks are moving on." It had been reported earlier in the day from Washington that reconnaisance photos had shown the "hurried" loading of the Soviet missiles, even aboard freighters not ordinarily equipped to carry them.)

Bases Reported Stripped. The Defense Department announced Nov. 8 that U.S. aerial surveillance of Cuba had shown that "all known" medium-range and intermediate-range ballistic missile bases in Cuba "have been dismantled" and that the Soviet missiles were being withdrawn aboard Russian vessels.

Partial text of the statement:

"The U.S. government has confirmed, through aerial reconnaissance, that medium-range ballistic missile and intermediate-range ballistic missile equipment is being removed from Cuba. Within the next 24 hours it expects to obtain additional confirmation through the close alongside observation of Soviet vessels by U.S. naval vessels. It is understood Soviet vessels will cooperate in this.

"As a result of aerial reconnaissance, the U.S. has photographs which indicate that all known MRBM [medium-range ballistic missile] and IRBM [intermediate-range ballistic missile] missile bases in Cuba have been dismantled."

"Photographs and visual inspection from U.S. naval vessels should provide further confirmation that the actual missiles (normally carried in missile transporters that have been photographed on board these vessels) have left Cuba.

"Intensive discussions are continuing with respect to the other provisions of the understanding between Pres. Kennedy and Chrmn. Khrushchev."

Inspection on High Seas

Ships Checked. The first inspection of Soviet ships removing missiles from Cuba was made by U.S. ships Nov. 8. Within 3 days, Soviet vessels carrying 42 MRBMs had passed through the U.S. blockade after submitting to the system of alongside scrutiny devised to assure the U.S. of the weapons' removal. None of the Soviet ships were boarded.

The beginning of the shipboard missile count was announced by Asst. Defense Secy. Arthur Sylvester in Washington Nov. 9. According to Sylvester, the first Soviet ship contacted, the *Alapayevsk*, had proved to be carrying only missile support and earth moving equipment. The first 3 Russian ships intercepted Nov. 9—the freighters *Dvinogorsk*, *Votgoles* and *Labinsk*—were found to be carrying missiles on deck; they cooperated in the alongside inspection by pulling off tarpaulins to permit the rockets to be photographed and examined from circling U.S. destroyers and helicopters. The missiles, without warheads and partially dismantled, were encased in tight weatherproof coverings and apparently were identifiable.

U.S. newsmen were permitted to observe the 6th interception of a Soviet missile ship from Navy patrol planes Nov. 10. They reported that the captain of the vessel, the *Ansonov*, had the tarpaulins removed from parts of 2 of the 8 missiles carried on his deck but refused to comply with requests from the U.S. destroyer *Barry* to uncover the rest of the missiles. The *Ansonov* reportedly signaled that

U.S. destroyer *Vesole* (below) overtakes Soviet freighter *Polzunov* in Atlantic Nov. 10 and inspects its deck cargo to verify Soviet removal of missiles.   (Wide World)

to do so was against instructions it had received from the Soviet government. The vessel was permitted to proceed despite its failure to comply with the request, and Pentagon officials said later Nov. 10 that the vessel's action had satisfied the minimal terms of the counting agreement.

Deputy Defense Secy. Roswell L. Gilpatric disclosed Nov. 11 that U.S. naval units had counted 42 MRBMs on the decks of Russian ships leaving Cuba. Gilpatric added, however, that the U.S. could not "be sure that 42 was the maximum number that the Soviets brought to Cuba" until on-site inspection had been carried out on the island. Gilpatric's disclosure was made in an interview broadcast by ABC-TV's "Issues and Answers" program. Although it was not initially regarded as an official statement, Pentagon spokesmen later in the day referred newsmen to the interview as the only information to be made available to the press on the Cuban situation that day.

Gilpatric's statement was the first official confirmation that the U.S. believed the 42 missiles reported sent to Cuba had been MRBMs only and did not include IRBMs.

Bombers' Withdrawal

U.S. Seeks Bombers' Recall. The Kennedy Administration took steps early in November to force the withdrawal of the IL-28 jet medium bombers shipped to Cuba by the Soviet Union.

(The IL-28 was the Russian equivalent of the U.S. B-47 bomber; in production since 1950, it had a top speed of 600 mph. and an operating radius of 700 to 750 miles. It was considered obsolete by current military standards.)

Amb. Stevenson and John J. McCloy took up the question of the IL-28s Nov. 4 at a UN meeting with Kuznetsov. They were said to have warned that refusal to withdraw the bombers would result in an intensification of the U.S. blockade. The Soviet representatives were reported to have stressed that the planes were obsolete and no threat to the U.S., but they were said to have conceded that they constituted "offensive weapons" within the meaning of the Oct. 28 Khrushchev-Kennedy accords.

McCloy was reported to have insisted that the U.S. took a "most serious view" of the bombers' presence in Cuba and could not give a formal pledge against invading the island until it had verification of the removal of both the Soviet missiles and bombers.

(Deputy Defense Secy. Gilpatric said Nov. 11 that the U.S. held "the Soviets responsible for the types of military equipment . . . furnished to Castro, and . . . we regard the removal of those bombers as within the capacity of the Soviets." Gilpatric was referring to Cuban arguments that the ownership of the bombers had been transferred to Cuba and that therefore the planes could not be removed by the USSR.)

Red Cross Rôle Dropped. U.S., Soviet and Cuban representatives continued their UN negotiations on terms for the inspection of Soviet weapons in Cuba despite the collapse Nov. 13 of talks on planned Red Cross inspection of Soviet cargoes bound for the island.

UN spokesmen in New York disclosed Nov. 13 that the talks on Red Cross inspection of Soviet shipping to Cuba had been recessed and further action on the proposal was "in abeyance." The abandonment of the proposal apparently was due to the inability of the U.S., Soviet and Cuban negotiators to agree on the processes and time limit to be applied to the Red Cross participation. It was reported that the U.S. had insisted that the Red Cross inspectors check all Cuba-bound shipping for at least 30 days; the USSR was said to have demanded that the inspection be limited to vessels of Soviet-bloc origin or charter and be held to a shorter period. The USSR also apparently had objected to the Red Cross insistence that its inspectors operate under UN authority.

The abortive negotiations for Red Cross participation in the envisaged inspection had been carried out in New York by Paul J. Reugger, former president of the International Committee of the Red Cross (ICRC), and Melchior Borsinger, secretary of the ICRC executive board. (The ICRC was an entirely Swiss organization.)

Cuba Warns Vs. Flights. Premier Castro warned Nov. 15 in a letter to U Thant that Cuba would shoot down any U.S. military planes entering its airspace for reconnaissance.

Castro's note, published by the UN Nov. 16, announced Cuba's "refusal to allow unilateral inspection by any body, national or international, on Cuban territory." It asserted that the U.S. already had verified the withdrawal of Soviet strategic missiles from Cuba by high-seas inspection. It charged that despite this verification, "every day the incursions of warplanes over our territory become more alarming; military aircraft harass our air bases, make low-level flights over our military defenses and photograph not only the dismantled strategic missile installations but . . . our entire territory foot by foot and inch by inch."

The Cuban message warned: "to the extent of the fire power of our anti-aircraft weapons, any warplane which violates the sovereignty of Cuba, by invading our airspace, can only do so at the risk of being destroyed."

USSR to Recall Bombers. It was announced Nov. 20 that Khrushchev had agreed to withdraw the IL-28 bombers and that the U.S. would reciprocate by ending its naval blockade of Cuba.

The announcement was made by Pres. Kennedy at a televised news conference in Washington. Mr. Kennedy disclosed that the accord had been worked out in his continuing exchange of secret letters with Khrushchev and that Khrushchev's message confirming the agreement had been received only "hours ago." The President made the announcement in a prepared statement read at the start of his news conference. It said in part:

"I have today been informed by Chrmn. Khrushchev that all of the IL-28 bombers now in Cuba will be withdrawn in 30 days. He also agreed that these planes can be observed and counted as they leave. Inasmuch as this goes a long way toward reducing the danger which faced this hemisphere 4 weeks ago, I have . . . instructed the Secretary of Defense to lift our naval quarantine."

"The evidence to date indicates that all known offensive missile sites in Cuba have been dismantled. The missiles and their associated equipment have been loaded on Soviet ships. And our inspection at sea of these departing ships has confirmed that the number of missiles reported by the Soviet Union as having been brought into Cuba, which closely corresponded to our own information, has now been removed.

". . . The Soviet government has stated that all nuclear weapons have been withdrawn from Cuba and no offensive weapons will be reintroduced."

"The Cuban government has not yet permitted the United Nations to verify whether all offensive weapons have been removed, and no lasting safeguards have yet been established against the future reintroduction of offensive weapons back into Cuba.

"Consequently, if the Western Hemisphere is to continue to be protected against offensive weapons, this government has no choice but to pursue its own means of checking on military activities in Cuba."[*]

"As for our part, if all offensive weapons systems are removed from Cuba and kept out of the hemisphere . . . under adequate verification and safeguards, and if Cuba is not used for the export of aggressive Communist purposes, there will be peace in the Caribbean. . . .

"We will not, of course, abandon the political, economic and other efforts . . . to halt subversion from Cuba. . . . But these policies are very different from any intent to launch a military invasion of the island."

Cuban Acquiescence. The announcement of the promised withdrawal of Soviet bombers from Cuba and of the lifting of the U.S. blockade came only hours after Castro had informed the UN that he would not object to the USSR's removal of the planes.

Castro's reversal of policy was attributed to pressure by Khrushchev and First Deputy Premier Mikoyan, the USSR's special envoy to Havana during the crisis.

Cuba's new position was made known in a note from Castro delivered to Acting Secy. Gen. Thant early Nov.

20. It charged the Kennedy Administration with trying to blame Cuba for blocking a peaceful settlement of the crisis and said that the U.S. had "made the IL-28 medium bombers stationed on Cuban territory the crux of the problem." "These planes," it declared, "are the property of the Soviet government. They were brought to Cuba for the defense of our country when faced with aggression. . . . They are antiquated. . . ."

The note asserted that it was "clear that the position of the . . . United States in demanding the withdrawal of these planes merely constitutes a pretext for maintaining tension, prolonging the crisis and continuing its policy of force. Nevertheless, if the Soviet government considers it desirable for the smooth conduct of the negotiations and the solution of the crisis to withdraw these planes, . . . Cuba will not object to this decision."

U.S. Ends Blockade

Quarantine Lifted. The U.S. naval quarantine of Cuba was ended Nov. 20. Defense Department spokesmen confirmed Nov. 21 that the Atlantic Fleet commander-in-chief had ordered the blockading vessels to discontinue the quarantine, effective at 6:45 p.m. the previous day. The 2 task forces involved in the blockade—Task Forces 136, made up of U.S. vessels, and 137, grouping the U.S. units and a

---

*Mr. Kennedy said: The U.S. had identified "a number of Soviet ground combat units in Cuba"; "although we are informed that these and other Soviet units were associated with the protection of offensive weapons systems and will also be withdrawn in due course, I repeat that we would like nothing better than adequate international arrangement for . . . verification in Cuba." (U.S. intelligence officials disclosed to Washington newsmen Nov. 1 that Cuban armed forces were known to have been equipped with "Frog" [firing range over ground] missiles capable of hurling conventional or tactical nuclear warheads 20 to 25 miles. These informants were said to have indicated that there was no evidence of the presence in Cuba of nuclear warheads for the weapons, which were truck-mounted. The missiles were said to fall within the category of defensive weapons, to which the U.S. had recognized Cuba's right. The U.S. was said to have superior missiles at the Guantánamo Naval Base.)

few Argentine ships—were dissolved and their vessels returned to normal duties. 63 vessels were on blockade duty off Cuba when the action was suspended.†

Defense Department officials disclosed Nov. 21 that units of Anti-Submarine Warfare, Atlantic, had been ordered to take up positions off Cuba to locate and photograph Soviet ships leaving the island with IL-28 bombers being returned to the USSR. The U.S. planned to maintain its aerial surveillance of Cuba for an indefinite period. It had been made clear by the U.S. that the photo-missions would be continued until a more adequate inspection system was devised to give assurance against the reintroduction of offensive weapons to Cuba.

(No information was made available on the number of vessels that had been halted, boarded or searched during the blockade, but the Defense Department had reported Nov. 14 that 49 vessels from Communist and non-Communist countries had been permitted to pass through the blockade to Cuban ports since the blockade started.)

USSR Ends Military Alert. The Soviet Union announced Nov. 21 that it had canceled the military alert ordered for the armed forces of the USSR and other Warsaw Treaty nations Oct. 23, at the height of the crisis. A government statement in Moscow said this action had been ordered in view of Pres. Kennedy's cancellation of the U.S. naval quarantine of Cuba.

The unusually detailed Moscow statement said that Marshal Rodion Y. Malinovsky, Soviet defense minister, had been ordered to have "rocket troops and intercontinental rocket troops transferred from full combat alert to normal training and activity." Similar measures were ordered for the USSR's "anti-aircraft rocket defense," fighter and "strategic aviation" forces, naval units and infantry and other ground forces. It said submarines had been ordered "to return to their regular stations." Military leaves and discharges, suspended at the start of the crisis, were ordered resumed.

A separate announcement said that Marshal Grechko, commander-in-chief of Warsaw Treaty forces, had issued similar orders for Soviet-bloc units under his command.

(The UPI reported from Havana Nov. 24 that Cuban militia units mobilized at the start of the missile base crisis apparently were being released. Truck convoys had begun returning militiamen to their homes all over Cuba Nov. 23. The official Cuban press and radio apparently confirmed the report Nov. 24 with the issuance of a propaganda call to the militiamen to return "from the trenches to the production front.")

JFK Inspects Forces. Pres. Kennedy Nov. 26 inspected U.S. military forces that had been alerted and deployed near the U.S.' southeast coast during the Cuban crisis. The President visited units at Ft. Stewart, Ga., Homestead AF Base, Fla. and the Boca Chica (Fla.) Naval Air Station.

Mr. Kennedy's inspection trip was undertaken after the U.S. had ordered the following steps Nov. 21 to relax the military combat alert, begun one month earlier: (a) release of the 14,-200 Air Force reservists recalled with the activation of 24 troop-carrier squadrons; (b) rescinding of the orders extending the active duty tours

---

†Reports that U.S. naval units involved in the Cuban blockade had detected and tracked Soviet submarines were confirmed Nov. 9 by Adm. George W. Anderson Jr., chief of naval operations. Anderson, addressing the N.Y. Council of the Navy League, said that "the presence of Russian submarines in Caribbean and Atlantic waters provided perhaps the first opportunity since World War II for our anti-submarine warfare forces to exercise at their trade." He added that "some of those submarines they detected did—after long periods of surveillance—come to the surface."

of Navy and Marine personnel whose enlistments expired after Oct. 23; (c) cancellation of the special alert imposed on combat forces of the Strategic Air Command; (d) redeployment of the interceptors of the Air Defense Command. (It was reported Nov. 21 that during the special alert 90% of SAC's 1,600 bombers had been placed on 15-minute ground alert and that the command's airborne alert—in which an estimated 30 bombers armed for a potential nuclear strike were aloft at all times—had been substantially expanded. The Defense Department announced Nov. 27 that the special airborne alert had been ended and that "a phased redeployment" of forces sent to the U.S.' southeast "has begun.")

Despite these steps to relax the U.S.' combat posture, substantial concentrations of ground, naval and air forces remained in the southeast.

Inspection in U.S. Demanded. Cuban Pres. Dorticós and Premier Castro informed U Thant in a memo delivered Nov. 27 that they would allow UN inspection of the dismantling of Soviet missile bases in Cuba only if similar inspection was made of bases in the U.S. at which, they said, anti-Castro forces were being prepared to invade Cuba.

The memo insisted that the UN "must verify, on the continental soil of the United States, Puerto Rico and other sites where aggression is being prepared against Cuba, the dismantling of training camps of mercenaries, spies, saboteurs and terrorists; centers where subversion is being prepared and bases from which pirate ships embark against our coasts."

The demand for inspection of U.S. territory was dismissed by U.S. State Department press chief Lincoln White Nov. 26 as "an attempt to avoid adequate international arrangements for inspection and verification that Cuba

will not once again become a base for offensive weapons."

Mikoyan Ends Mission. Soviet First Deputy Premier Mikoyan, sent to Cuba by Khrushchev at the height of the crisis, returned to Moscow Dec. 2 after completing his Havana talks and stopping in the U.S. for meetings with Pres. Kennedy and other Administration leaders.

Mikoyan ended his 24-day Cuban mission Nov. 26. Addressing the Cuban people on TV on the eve of his departure, he asserted that "the cause of Cuba is the cause of the Soviet Union," which would remain the Castro régime's "friend and protector." Defending the withdrawal of the Soviet missiles, he declared this had been done to prevent nuclear war and to help "the Cuban people to achieve their independence."

Mikoyan flew to New York Nov. 26 for talks with U Thant and then went to Washington, where he conferred for 3¼ hours Nov. 29 with Pres. Kennedy and State Secy. Rusk at the White House. Speaking to newsmen Dec. 1 before his departure from Andrews Air Force Base, Md., he declared that "in our conversations in Washington we became convinced of goodwill on the American side and the same will be evinced by us."

Bombers Leave Cuba. 42 Soviet IL-28 bombers left Cuba aboard Russian-bound ships Dec. 1-6. The bombers, dismantled and crated on the vessels' decks, were counted by observers aboard U.S. planes as they began the return trip to the USSR. The 42 planes were believed to be all of the Soviet bombers sent to Cuba.

The withdrawal was disclosed by a U.S. Defense Department statement issued Dec. 3. The statement said: "IL-28 aircraft that have been seen on the island of Cuba are in the process of being withdrawn. Photo-

graphs indicate that the Soviet ship
Okhotsk was proceeding on Dec. 1
off the north coast of Cuba with 3
fuselages visible on the deck."

The completion of the withdrawal
was reported in 2 additional Defense
Department statements made public
Dec. 6. The statements said that the
42 planes had been shipped from
Mariel and Nuevitas, ports on Cuba's
north coast, and had been "voluntarily" displayed to circling U.S.
planes by the crews of the 3 Russian
ships involved. Aerial surveillance of
the San Julian and Holguin airfields,
at which most of the bombers had
been concentrated, had shown "no
IL-28 fuselages or crates remaining."

Russians Build Cuba Defense.
Washington officials reported Dec. 6-7
that intelligence information received
from anti-Castro sources in Cuba had
described intensive work by Soviet
military personnel on Cuban defenses.

The information, said to have been
dated Nov. 29, indicated that some
Soviet military camps had been closed
but that many Soviet military and
technical units remained despite the
departure of Russian missiles. The
Russians were said to be concentrat-
ing their efforts on construction of
bunker-type defenses, underground
fuel and munitions depots, airstrips
and fighter defense facilities and anti-
invasion defenses. Cuba's air force,
believed to include 150 MiG-15s and
some MiG-21 jet fighters, was said to
have been dispersed to prevent its
destruction in a sudden attack. The
short-range ground-to-ground and
ground-to-air missiles known to have
been sent to Cuba by the Russians
were said to have been emplaced in
fortified and underground centers.

U.S. State Secy. Rusk said at a
Washington news conference Dec. 10
that there seemed to be an "out
movement" of Soviet military personnel from Cuba, but the U.S. had only

an "inexact indication" of the extent
of the withdrawal. He declared that
the Western Hemisphere could not
accept as settled a situation in which
there remained "any Soviet military
presence in Cuba."

No-Invasion   Pledge   Demanded.
Khrushchev called on the U.S. Dec.
12 to honor its agreement not to invade Cuba as the USSR had honored
its promise to withdraw its missiles,
jet bombers and other aggressive
weapons from the island.

Khrushchev's demand was made in
a 2½-hour speech before the USSR's
Supreme Soviet (Parliament) in Moscow. In the address he defended his
Cuban policies on the ground that
they had turned the recent crisis away
from a nuclear war that would have
destroyed the world.

The USSR, Khrushchev declared,
would continue to fulfill its share of
the Cuban agreement "so long as the
other side stands by this understanding." "But," he added, "if the commitments assumed are not observed by
the other side, we shall be compelled
to take . . . action."

Khrushchev acknowledged that the
USSR had sent rockets to Cuba and
had made them "ready for launching,
ready for action." But he denied that
these missiles had been intended to
be used for an attack on the U.S.
"Had we wanted to start war against
the United States," he declared, "we
would not have agreed to dismantle
the rockets. . . . We would have
brought them into play. But we did
not do that because we had no such
aims." Khrushchev's assertion that the
Soviet withdrawal of weapons from
Cuba had saved the world from potential destruction was cheered loudly
by the 1,443 deputies present.

(Pres. Kennedy was questioned by
reporters Dec. 12 as to whether, in
view of the Khrushchev speech, the
U.S. would make a formal pledge not

to invade Cuba without first having obtained the inspection it sought of Cuban military facilities. He declined to comment on the Khrushchev speech, but he said that the U.S. was continuing its UN negotiations with the Russians on "the question of weapons, inspection, aerial observance, invasion," and he was hopeful they would succeed.)

U.S. Administration's Actions Debated

JFK Supported, Criticized. U.S. political leaders of both parties generally supported Pres. Kennedy's decision to blockade Cuba, but many Republicans criticized the President's actions on 2 points: (1) They said their party had been warning of a Soviet missile buildup in Cuba and had been urging strong action long before Mr. Kennedy acted; (2) they suggested that the President might have timed his move to gain advantage in the Nov. 6 elections.

Pres. Kennedy had called 19 Congressional leaders of both parties to Washington Oct. 22 to brief them on the situation before his nationwide TV address. The leaders were called back to the White House Oct. 24 for a 2d briefing. Although both briefings were secret, a leak (later virtually confirmed) revealed that Chrmn. Richard B. Russell (D., Ga.) of the Senate Armed Forces Committee had proposed at the Oct. 22 briefing that the U.S. invade Cuba immediately instead of awaiting the outcome of a blockade. Russell was reported to have said that the quarantine did not rule out a possible invasion later and that he thought there would be fewer casualties if the U.S. invaded immediately. Pres. Kennedy was reported to have opposed an immediate invasion on the ground that it would mean largescale fighting before the USSR had a chance to decide what

to do about the U.S. challenge.

Pres. Kennedy Oct. 24 invited every member of Congress to attend closed-door briefings on the situation. Since most Congress members were campaigning in their home states, the briefings were scheduled for 5 cities across the country—New York, Chicago, Atlanta, Fort Worth and San Francisco.

Ex-Pres. Dwight D. Eisenhower, kept up-to-date on the crisis by the Kennedy Administration, continued his vigorous campaigning on behalf of GOP candidates and his denunciations of the Kennedy "clique." Speaking at a Republican rally in Pittsburgh's Civic Arena Oct. 25, Eisenhower said that Americans must make every personal sacrifice needed in the crisis but that "no foreign crisis excuses [us] from our continuing and lifelong responsibility as citizens to vote our convictions about the sort of government we want. . . ."

Mr. Kennedy phoned ex-Presidents Eisenhower, Harry S. Truman and Herbert Hoover Oct. 28 to tell them personally of Khrushchev's letter revealing that he had ordered the dismantling of Soviet missiles in Cuba and to consult them before replying. Truman told reporters later Oct. 28: "I knew very well if we met Khrushchev head on that's what he'd do. Kennedy did meet them head on and that's what counted."

Ex-Vice Pres. Richard M. Nixon, campaigning as GOP gubernatorial candidate in California, asserted in Fremont Oct. 26 that Pres. Kennedy's action had decreased the threat of war. If missiles remained in Cuba, Nixon said, Cuban Premier Castro, whom Nixon described as "a madman," might eventually win control of them and use them for wide destruction. Nixon said in Los Angeles Oct. 28: "This demonstrates again that when you stand up to Communist aggressors, they back down."

Sen. John J. Sparkman (D., Ala.) disagreed Oct. 25 with complaints that Mr. Kennedy should have acted sooner. Speaking in Atlanta after one of the regional State Department briefings for Congress members and governors, Sparkman said: "It was hard to resist the pressure [for early action against Cuba] . . . but we needed the support of other countries, particularly those of the Western Hemisphere. We got it by working with these nations over the past 6 weeks. . . ."

Sen. Homer E. Capehart (R., Ind.) asserted Oct. 28: "I have said in 100 speeches that if we were firm and would blockade and take military action against Cuba, Khrushchev would put his tail between his legs and run, and that's what he is doing."

Sen. Barry Goldwater (R., Ariz.) called on Mr. Kennedy Nov. 5 to repudiate the U.S. implied pledge not to invade Cuba. Goldwater, in a statement issued jointly with Rep. Bob Wilson (R., Calif.) (the 2 were the respective chairmen of the Senate and House Republican campaign committees), declared that the no-invasion pledge had "locked Castro and communism into Latin America and thrown away the key to their removal." The statement called this pledge the U.S.' "most serious foreign policy error since the Bay of Pigs fiasco last year."

Stevenson Dispute. A major controversy became apparent in Kennedy Administration circles Dec. 3 over allegations concerning U.S. Amb.-to-UN Stevenson's rôle in the high-level deliberations leading to Pres. Kennedy's decision to blockade Cuba.

The allegations were made in an article written for the Dec. 8 issue of the *Saturday Evening Post* by Stewart Alsop, *Post* Washington correspondent, and Charles Bartlett, Washington reporter of the *Chattanooga* (Tenn.) *Times*. Both writers were considered friends of the President. Bartlett, described as a close personal confidante of Mr. Kennedy, had been the first newsman to report the policy disagreement that led to the replacement in 1961 of State Undersecy. Chester W. Bowles.

The *Post* article purported to give an account of the secret deliberations held on the crisis by the Executive Committee of the National Security Council, a supreme Presidential advisory body formed at the start of the missile affair.° The article asserted that Stevenson had "dissented" strongly from the blockade decision reached by the committee.

It said: "Only Adlai Stevenson . . . dissented from the [blockade] consensus. There is disagreement in retrospect about what Stevenson really wanted. 'Adlai wanted a Munich,' says a nonadmiring official who learned of his proposal. 'He wanted to trade the Turkish, Italian and British missile bases for the Cuban bases.' . . . There seems to be no doubt that he preferred political negotiation to the alternative of military action. White House aide Arthur Schlesinger was assigned to write the uncompromising speech which Stevenson delivered at the UN on Tuesday [Oct. 23], and tough-minded John McCloy was summoned from a business conference in Germany to work with Stevenson in the UN negotiations."

*Formation of the NSC Executive Committee had been announced by the White House Oct. 23. The committee, which numbered 12 or more members, 7 more than the full NSC, was viewed as a formalization of an "inner group" of Administration officials that the President relied on for policy formation in crisis situations. Committee members: Stevenson, Atty. Gen. Robert F. Kennedy, Defense Secy. Robert S. McNamara, McGeorge Bundy (special Presidential assistant for national security affairs), Vice President Johnson, State Secy. Dean Rusk, Treasury Secy. Douglas Dillon, Gen. Maxwell D. Taylor (chairman of the Joint Chiefs of Staff), CIA Dir. John A. McCone, State Under-secy. George W. Ball, Deputy Defense Secy. Roswell L. Gilpatric, Amb.-at-Large Llewellyn E. Thompson Jr. and Theodore Sorenson, special counsel to the President. Other persons, among them John J. McCloy, in charge of UN negotiations on the Cuba question, took part in committee meetings as required.

The article's allegations, made known in New York Dec. 2, were denounced as untrue by Stevenson Dec. 3. Speaking with reporters at the State Department, after meeting privately with Pres. Kennedy during a visit to the White House for an NSC session, Stevenson declared that the article was "inaccurate and grossly misrepresenting" of his views. "The facts are," he said, "that I never proposed to trade our bases abroad to get the Soviets out of their missile bases in Cuba, and that I favored the blockade on the shipment of more Soviet arms to Cuba." He declined further comment on his position in the pre-blockade deliberations on the ground that advice given the President in such conditions was secret.

Pres. Kennedy expressed his "fullest confidence" in Stevenson in a personal letter to Stevenson dated Dec. 4 and made public Dec. 5. Mr. Kennedy's letter, addressed to "Dear Adlai," said: "This is just a note to tell you again how deeply I regret the unfortunate stir which has arisen over the statements contained in the *Saturday Evening Post* article. I think you know how greatly we have all admired your performance at the United Nations. . . . I have, of course, valued your advice very highly. That we have eliminated the nuclear menace from Cuba is the best evidence of the prudence of our policy and its execution, in which you played such an active part. . . . Your continued work at the United Nations will be of inestimable value. . . ."

Stevenson, interviewed Dec. 5 on NBC-TV's "Today" program, declared that the *Post* article "was wrong in literally every detail" and again denied that he had opposed blockading Cuba. Stevenson, although he again refused to reveal details of his position in the NSC deliberations, said that he had "emphatically" favored the use of a blockade against Cuba

3 days before the final decision was taken. Stevenson compared the *Post* article to a form of character assassination. "I am used to assassins," he said. "I remember [the late Sen. Joseph R.] McCarthy very well indeed."

Pres. Kennedy, at a press conference Dec. 12, lauded Stevenson's "distinguished service" but did not specifically answer a query as to the truth of the *Saturday Evening Post* article. Mr. Kennedy said: "It is my view that when the final consensus was reached and when I finally made a judgment . . . that every member of the Executive Committee of the National Security Council supported the policy we finally adopted." Asked to comment on a *Life* magazine article (Dec. 14 edition) implicating the White House in the preparation of the Alsop-Bartlett story, Mr. Kennedy said that 2 points were involved: "One was whether the White House had in any way authorized or suggested the article in the *Saturday Evening Post,* and, No. 2, whether the White House had made members of the National Security Council available. Both of those were untrue."

Doubts Voiced. Correspondents in Washington and London reported that military and political observers had found many faults with the Kennedy Administration's story of how and when the Cuban missile bases were discovered and of what actually was discovered. Among reasons given for their doubts (and some Administration answers):

▶ Cuban refugee reports that the Russians were setting up offensive missile bases in Cuba were denied by the Kennedy Administration at first, just as other reports by refugees— about the arrival in Cuba of Soviet arms and military technicians and the use of the technicians to build bases— were first denied but later confirmed by the Administration. The Administration also had denied a charge made by Sen. Kenneth B. Keating (R., N.Y.) in the Senate Oct. 10 that "construction has begun on at least a half-dozen launching sites for intermediate-range tactical missiles."

Administration spokesmen said Oct. 26 that all refugee reports were carefully checked but were usually found to be highly inflated. Short-range surface-to-air anti-aircraft missiles were described by refugees as intercontinental ballistic missiles, the spokesmen said, and similar exaggerations were proven in other refugee reports.* According to Defense Secy. McNamara, the first "hard" evidence of surface-to-surface missile bases was in aerial reconnaissance photos made Oct. 15, shown to McNamara that same night and shown to Pres. Kennedy Oct. 16. The construction had not been photographed before this because the bases were built with amazing speed—in some cases literally overnight, in others, according to McNamara, "in a matter of days."

▶ Pres. Kennedy and his military aides described the offensive weapons in Cuba as medium-range missiles with a range of 1,200 miles and intermediate-range missiles with a range of 2,400 miles. Photos and other evidence cited by the doubters, including the military correspondent of the London *Times*, indicated that the

medium-range missiles were the Soviet T-1, generally credited with a range of only 400 to 600 (possibly 700) miles, and that the intermediate-range missiles were the T-2 and/or T-4, the former generally credited with a range of no more than 1,500 miles, the latter with a range of 1,000 miles.

According to Administration officials, only medium-range missiles were in firing position in Cuba, and none of the sites for the intermediate-range missiles were even near completion. Members of Congress who attended a State Department briefing on the crisis in Chicago Oct. 25 said they were told that the sites for the latter, when finished, would be capable of firing 32 to 40 2,000-mile-range missiles simultaneously.

News Curbs. Kennedy Administration control of the flow of information during the Cuban crisis gave rise to charges that officials had restricted, manipulated and suppressed news to mislead public opinion.

A Defense Department memo sent to newspaper editors and radio and TV news directors Oct. 24 had urged them to use extreme discretion in reporting information falling within 12 areas of news lest their reports impair U.S. security. A 2d memo, issued Oct. 27 by Asst. Defense Secy. Sylvester, forbade any department official to give information to the press without specific Pentagon clearance. A similar State Department memo, issued by Asst. State Secy. Robert J. Manning, ordered all department officials to report to him immediately any conversations held with reporters.

Responding to newsmen's protests that they had been mislead and cut off from legitimate information, Sylvester asserted Oct. 30 that management of news was "part of the arsenal of weaponry" available to the Administration in time of crisis. He was

---

*José Antonio Lanuza, spokesman for the exile Cuban Student Directorate, said in Miami Oct. 23 that his organization had pinpointed at least 15 missile bases already built or under construction in Cuba. He said that submarine bases had also been reported under construction in Mariel Bay and in Siguanea Bay on the Isle of Pines and that 8 missile-firing Soviet submarines were on permanent duty in Mariel Bay. Lanuza said Nov. 9 that according to reports received from agents in Cuba, the USSR had sent 80 missiles to Cuba and had aided the construction of underground hangars and repair facilities for MiG fighters and IL-28 bombers at 10 Cuban airfields. Luis Fernández Roche, leader of the directorate, said Nov. 12 in a New York radio interview: 25 underground missile-launching silos had been built in Cuba by Soviet technicians; they had been maintained with missiles despite the Soviet shipments from the island; the Soviets "still have missiles, they still have underground bases, they still have everything" in Cuba.

quoted as saying: "in the kind of world we live in, the generation of news by . . . government becomes one weapon in a strained situation; the results, in my opinion, justify the methods we used."

Pres. Kennedy commented Nov. 20 that despite newsmen's protests, "it seems to me that the papers more or less reflected quite accurately the state of our negotiations with the Soviet Union" during the crisis. He said that the State Department directive would be withdrawn but that the Defense Department memo would stand because it had not been proven that "the flow of essential news of the Pentagon" had been restricted by it. He agreed with reporters, however, that aside from "highly sensitive information," news of government activities should "pour out" to the public.

The Defense Department Nov. 23 canceled the order barring personnel from talking to reporters about matters in the 12 security-sensitive areas but kept in force the order requiring personnel to report any conversations with reporters. The similar State Department order was canceled Nov. 27.

Sylvester, in reply Nov. 27 to a Nov. 17 National Editorial Association resolution protesting Pentagon news curbs, denied that legitimate news had been curtailed. "By any measure you choose to use, the news flow has increased here," he asserted. Sylvester reiterated in a talk before the Deadline Club in New York Dec. 6 that there had been no lessening of the amount of news from the Pentagon. He was reported to have said that "the inherent right of the government to lie to save itself when faced with nuclear disaster is basic."

Report to Nation. Pres. Kennedy gave the American people a personal report Dec. 17 on his Administration's first 2 years.

The President's report was made in an hour-long unrehearsed interview filmed Dec. 16 in his White House office and televised the next day. The interview was conducted by William H. Lawrence of ABC, George Herman of CBS and Sander Vanocur of NBC.

Major points made by Mr. Kennedy:

Peace—There was no reason why, ultimately, the U.S. and USSR should not be able to live in peace with each other. "If Mr. Khrushchev would concern himself with the real interests of the people of the Soviet Union, . . . there is no real reason why the United States and the Soviet Union, separated by many thousands of miles of land and water . . . should not be able to live in peace."

The major obstacle to stability between the 2 blocs was the Communist side's "constant desires to change the balance of power in the world." "In fairness" to Khrushchev, the West recognized that "we are better off with the Khrushchev view than we are with the Chinese Communist view. . . . But Mr. Khrushchev does not wish us well, unfortunately."

There was little evidence that the U.S. would be able to "come to any real understanding" with Khrushchev soon. The U.S. did not view a new summit conference as useful until "we see some breakthrough in some one area," such as disarmament or a ban on nuclear testing.

Cuban crisis—"The Cuban effort has made it more difficult for us to carry out any successful negotiations because this was an effort to materially change the balance of power. It was done in secret, [and] steps were taken really to deceive us by every means. . . ."

The Soviet attempt to emplace missiles in Cuba was an "imprudent" act that showed "the Communists have a completely twisted view of the

United States" and its readiness to defend itself. The USSR had planned to reveal its Cuban missile base after completion, in November, but was forced to retreat by the U.S.' resolute action.

There was little hope that the U.S. ever would obtain on-site inspection of Cuban bases because "a totalitarian system cannot accept the kind of inspection which really is desirable." If Cuba accepted such inspection this would be a step toward a similar opening of the USSR; "they are not going to open it," and "the camera, I think, is actually to be our best inspector."

"If we had had to act . . . in the first 24 hours [after learning of the Soviet missiles in Cuba], I don't think, probably, we would have chosen as prudently as we finally did —a quarantine against the use of offensive weapons."

A "general consensus" on the Cuban quarantine emerged only "after all alternatives were examined" and some disagreement had been expressed by Presidential advisers. "The course we finally adopted had the advantage of permitting other steps if this one was unsuccessful. . . . We were starting . . . at a minimum place. Then, if that were unsuccessful, we could have gradually stepped it up until we had gone into a much more massive action. . . ."

The advice supporting the blockade decision "was the right one." The advice supporting the Apr. 1961 decision to back the Bay of Pigs invaders "was wrong." "And I was responsible."

War—The Cuban crisis was a lesson to the USSR in the U.S.' readiness to defend itself. The danger remained, however, that the USSR might mistakenly assess the U.S.' intentions in some future crisis, and

"one mistake can make this whole thing blow up."

"Once he [Khrushchev] fires his missiles, it is all over anyway, because we are going to have sufficient resources to fire back at him and destroy the Soviet Union. When that day comes, and there is a massive [missile] exchange, then that is the end, because you are talking about Western Europe, the Soviet Union, the United States, of 150 million casualties in the first 18 hours."

Communist Policy Differences

Reports from Moscow, Peiping and other Soviet-bloc capitals in October indicated that China had opposed the USSR's decision to withdraw its missiles from Cuba and dismantle its launching sites there. Chinese leaders criticized the Russian action as a "retreat" in the face of U.S. imperialism.

The Chinese charges and the Russian defense against them, were aired in government and press statements and in addresses delivered by representatives of both sides at a series of Communist party congresses held in the Soviet bloc and Western Europe. Although the language of the initial Chinese-Russian exchange was elliptical and rarely referred specifically to the matter—or the persons—at question, the series of strong Chinese attacks on policies identified personally with Khrushchev eventually forced the Russian leader to label China publicly as the opponent of his efforts to prevent nuclear war.

China Vs. 'Appeasement.' The Chinese opposition to the USSR's handling of the Cuban crisis first become apparent after Khrushchev Oct. 28 had agreed to Pres. Kennedy's demand for the withdrawal of Russian missiles from Cuba.

The Chinese CP newspaper *Jenmin Jih Pao* (*Peiping Peoples Daily*) reported the Khrushchev-Kennedy agreement in a single paragraph on an inside page Oct. 29 but devoted the front page of the same edition to a Peiping mass rally that pledged Cuba China's support until it achieved "final, complete and thorough victory" against the U.S. The newspaper Oct. 31 published editorials in which it was implied that Khrushchev had erred in accepting the Kennedy demands. It declared that "the so-called assurance that the United States will not invade Cuba is nothing but a hoax." It contrasted Russia's attitude with China's full support of Premier Fidel Castro's revolutionary régime.

Hundreds of thousands of persons marched to the Cuban embassy in Peiping Nov. 3-5 to demonstrate support for Cuba. *Jenmin Jih Pao*, in a front-page editorial Nov. 5, condemned any "appeasement" of the U.S. in the Cuban situation. In an implied rebuke to Khrushchev, it declared: "To compromise with or meet the Kennedy government's truculent demands can only encourage the aggressor and will in no way insure world peace."

The USSR's alleged retreat in Cuba was attacked Nov. 8 by the Chinese chief delegate to the Bulgarian CP Congress in Sofia. Wu Hsiu-chuan, a member of the Chinese CP Central Committee, assailed the USSR for its Cuban failure and for its policies toward Yugoslavia and Albania. Wu reiterated China's support for the Castro régime and its demands for the U.S.' Guantánamo base and a halt to sanctions against Cuba as "the minimum that must be sought."

Articles appearing in the Chinese press and reported from Hong Kong Nov. 15-16 denounced the Cuban position taken by the "modern revisionists," a term considered an allusion both to Khrushchev and Yugoslav Pres. Tito. They charged that these interests, frightened by the confrontation with U.S. power in the Caribbean, had betrayed Marxist principles by withdrawing Soviet missile power from Cuba. *Jenmin Jih Pao* declared in a front-page editorial Nov. 18 that it was "nonsense" to claim that "peace had been saved" only by the withdrawal from Cuba.

USSR Gathers Support. Moscow dispatches made it clear Oct. 30-Nov. 10 that Khrushchev had turned to East European Communist leaders for support in his handling of the Cuban crisis and his rift with the Chinese. Nearly every East European Communist leader visited Moscow during that period, and all of them conferred secretly with Khrushchev on the Cuban problem. The Moscow visitors included Czech Pres. Antonin Novotny, who arrived Oct. 30, East German CP First Secy. Walter Ulbricht, who arrived Nov. 1, Rumanian CP First Secy. Gheorghe Gheorghiu-Dej, Bulgarian CP First Secy. Todor Zhivkov, Polish CP First Secy. Wladyslaw Gomulka, who arrived Nov. 3, and Hungarian Premier János Kádár, who was the last to visit Khrushchev and who left Moscow Nov. 10.

The traditional Kremlin address on the eve of the anniversary of the Bolshevik revolution, delivered Nov. 6 by Soviet First Deputy Premier Aleksei N. Kosygin, contained an explicit defense of the Soviet position in the Cuban crisis. Speaking to an audience of 8,000 Soviet and foreign dignitaries, Kosygin asked: "To whom is it not clear that if war had broken out it would have been a thermonuclear world war from the very outset?" Examining the question of "who made concessions to whom," Kosygin said: "We feel that there were concessions both from one and from the other side. They were concessions to sanity and peace. Some may ask whether it was worth making concessions. We feel that this compromise was in the interest of all nations, because on its basis it was possible to eliminate the threat of a thermonuclear world war."

The Soviet CP Nov. 18 denounced Communist critics of Khrushchev's Cuban policies for "pushing mankind toward thermonuclear war." CP Central Committee Secy. Boris N. Ponomarev, writing in *Pravda*, declared that Khrushchev's actions had been subjected to "unlimited slander" from critics within the Communist bloc. Ponomarev, considered a specialist in relations among the Communist parties, said that "neither bourgeois propagandists nor other falsifiers" could detract from Khrushchev's contribution to peace during the Cuban crisis. In a reference directed at Albanian leaders but presumed to apply to China, he charged that "they have undertaken an especially shameful and . . . provocative campaign in connection with the crisis in the Caribbean."

Controversy Debated. The Chinese-Russian dispute was brought before the annual congresses of the Hungarian, Italian and Czech Communist parties in late November and early December. Major developments in the controversy reported from the party congresses:

Hungarian congress—CP First Secy. Kádár, opening the Budapest meeting Nov. 21, declared that "we profoundly condemn the anti-partyite [Albanian CP First Secy.] Enver Hoxha and . . . all the dogmatic sectarians . . . making their appearance in other parties of the international Communist movement." Kádár asserted that Communist-bloc unity meant not only a unity of ideals "but also common action in the international sphere on the paramount questions of the revolutionary movement."

The Kádár attack on Albania was criticized Nov. 21 by Wu Hsiu-chuan, leader of the Chinese delegation to the Budapest meeting. Wu insisted, however, that China sought only fraternal ties with all Communist parties, especially that of the USSR.

Italian congress—China's position was attacked Dec. 3 by Soviet CP Secy. Frol R. Kozlov in an address at the Italian CP's 10th annual congress in Rome. Kozlov denounced the opponents of Soviet coexistence policies for adopting "an adventurist position which has nothing to do with Marxism." He charged that "such a position, even through camouflaged with pseudo-revolutionary phrases, becomes particularly harmful and dangerous" in an age of nuclear armaments.

The Chinese delegate to the Rome meeting, Chao Yi-ming, conceded in an address Dec. 4 that "differences" existed between the Chinese and Soviet parties. Referring to Communist-bloc foreign policy, he declared that "the nature of imperialism will not change nor will the policy of aggression." "Peace-loving peoples . . . must reply to aggression blow by blow," he said.

Czech congress—CP First Secy. Novotny opened the Czech party's 12th congress in Prague Dec. 4 with a speech in which he accused Albania, and, by implication, China, of having aided U.S. "imperialism" by attacks on the USSR's Cuban policies. Making clear his reference to the Chinese, Novotny declared that "those who listen to the Albanian leaders or support them in their aggression against the Communist movement put themselves in opposition to the declarations of the world's Communist parties." CP Central Committee Secy. Vladimir Koucky expressed regret Dec. 7 that the Chinese delegation, led by Wu Hsiu-chuan, had failed to join in criticism that Albanian leaders had voiced "disgusting slanders" against the USSR.

Khrushchev Counterattack. Khrushchev personally defended his actions during the Cuban missile crisis in a major foreign policy address broadcast Dec. 12 from the USSR's Supreme Soviet (parliament) in Moscow.

Khrushchev specifically identified China as the source of Communist bloc charges that he had "appeased" imperialism by his settlement of the Cuban crisis. It was believed to be the first official confirmation to the Russian people of Chinese opposition to Khrushchev's foreign policies.

Khrushchev made this rejoinder to the Chinese gibe that imperialism was a "paper tiger:" Let them remember that the "paper tiger has nuclear teeth" and "may still use them and should not be treated lightly."

Conceding that elements within the Communist bloc had opposed as a "retreat" his decision to accept U.S. demands for withdrawal of Soviet missiles and bombers from Cuba, Khrushchev declared: "In what way have we retreated, one may ask. Socialist Cuba exists. Cuba remains a beacon of Marxist-Leninist ideas in the Western Hemisphere. The impact of her revolutionary example will grow. The . . . United States has given a pledge not to invade Cuba. The threat of thermonuclear war has been averted. Is this a retreat?"

Khrushchev added: "Will anybody censure the People's Republic of China because fragments of colonialism remain intact [in China]. It woud be wrong to push China to any actions which she regards as untimely. If . . . China tolerates Macao and Hong Kong there must be weighty reasons for this. It would be ridiculous to use this for an accusation that they are making a concession to the British and Portuguese colonialists, that this is appeasement."

Chinese Demand Parley. Peiping radio announced Dec. 14 that Communist China had proposed a meeting of the world's Communist parties "in order to clarify what is right from what is wrong, to strengthen unity and to stand together against the enemy."

The demand for a world CP congress on the problem was repeated in an article published in *Jenmin Jih Pao* Dec. 15. The article, reputedly written by Mao Tse-tung, said that the USSR had broken the 1960 Moscow agreement against the public airing of Communist policy differences and had made it necessary to call another such meeting to reestablish Communist unity. Although not mentioning the USSR by name, the article criticized Moscow for "capitulationism" in withdrawing its missiles from Cuba and for "adventurism" for placing them there in the first place. The article added: "The whole world sees that we [China] neither requested the introduction of nuclear weapons in Cuba nor obstructed [their] withdrawal. . . . For us, therefore, there is no question whatsoever of 'adventurism,' still less of plunging the whole world into a thermonuclear war."

Khrushchev Assailed. Chinese leaders proclaimed their rejection of Khrushchev's policies again in a year-end statement published in *Jenmin Jih Pao* Dec. 31. The statement was considered to be China's reply to Khrushchev's Dec. 12 speech defending his actions in Cuba.

Ostensibly written to answer attacks made on Chinese views by Italian CP Secy. Gen. Palmiro Togliatti, the statement repeatedly referred to Togliatti and "other comrades" or "certain others." It was clear from the context—and from the fact that Khrushchev's remarks were quoted and criticized although not attributed to him—that these "others" were Khrushchev.

Points made by the Chinese statement, entitled "The Differences Between Comrade Togliatti and Us":

Cuba—"Those who accuse China of opposing peaceful coexistence also attack the Chinese people for supporting the just stand of the Cuban people in their struggle against U.S. imperialism." "Was there anything wrong in that? Yet some people have repeatedly charged China with creating difficulties in the Caribbean situation and with wanting to plunge the world into a thermonuclear war."

"Does this mean that China, also, should have applied pressure on Cuba to force it to accept international inspection, and that only by so doing would China have conformed to this so-called 'peaceful coexistence'? If there are people who give verbal support to Cuba's 5 demands [for settlement of the recent missile base crisis] but are actually opposed to the Chinese people's support for Cuba, are they not merely exposing the hypocrisy of their own support for Cuba's 5 demands?"

"The Chinese . . . have always maintained that the course of history is decided by the great strength of the masses of the people and not by any weapons. On more than one occasion we have made it clear that we neither called for the establishment of missile bases in Cuba nor obstructed the withdrawal of the so-called 'offensive weapons' from Cuba. We have never considered that it was a Marxist-Leninist attitude to brandish nuclear weapons as a way of settling international disputes. . . . What we did strongly oppose . . . is the sacrifice of another country's sovereignty as a means of reaching a compromise with imperialism. A compromise of this sort can only be regarded as 100% appeasement, a 'Munich' pure and simple."

Coexistence & imperialism—"Even more absurd is the allegation that 'a world without war' can be achieved through peaceful coexistence. In the present situation, it is possible to prevent imperialism from launching a new world war if all the peace-loving forces of the world unite . . . and fight together. But it is one thing to prevent a world war and another to eliminate all wars.

"Imperialism and the reactionaries are the source of war. . . . Oppressed nations and oppressed people are bound to rise in revolution. When imperialism and the reactionaries employ armed forces to suppress revolution, it is inevitable that civil wars and national liberation wars will occur. Marxist-Leninists have always maintained that only after the imperialist system has been overthrown . . . will it be possible to . . . reach 'a world without war'."

"The principle of peaceful coexistence can apply only to relations between countries with different social systems, not to relations between oppressed and oppressor nations, nor to relations between oppressed and oppressing classes."

Nuclear war—"Togliatti and certain others believe that the emergence of nuclear weapons 'has changed the nature of war.' . . . Actually, they believe that war is no longer the continuation of politics and that there is no longer any distinction between just and unjust wars. Thus they completely deny the fundamental Marxist-Leninist theory on war and peace.

"We hold that the emergence of nuclear weapons has not changed and cannot change the fundamental Marxist-Leninist theory with regard to war and peace. . . ."

"In practice, those who hold there is no longer any distinction between just and unjust war either oppose just wars or refuse to support them, and they have lapsed into the position of bourgeois pacifism, which is opposed to all wars."

"Togliatti and certain others talk volubly about 'the suicide of mankind' and the 'total destruction of mankind.' . . ."

"If, after we have done everything possible to prevent a nuclear war, imperialism should nevertheless unleash nuclear war, . . . it would result in the extinction of imperialism and definitely not in the extinction of mankind. The Moscow Statement [of 1960] points out that 'should the imperialist maniacs start war, the peoples will sweep capitalism out of existence and bury it'."

Tito Visits USSR. Yugoslav Pres. Tito visited the Soviet Union Dec. 3-21. His trip was described officially as a "holiday," but the Belgrade newspaper *Borba* said its purpose was to exchange views with Premier Khrushchev "on pressing international issues, on mutual relations between the 2 countries and on questions of mutual interest." (Tito's trip to Russia was made in response to an invitation delivered by Soviet Pres. Leonid Brezhnev and *Izvestia* editor Aleksei I. Adzhubei, Khrushchev's son-in-law, during a visit to Yugoslavia Sept. 24-Oct. 4.)

Tito, in an address to the Supreme Soviet Dec. 13, praised Khrushchev's policy during the Cuban missile crisis. He said the Soviet premier had acted "bravely at the most critical moment, taking into account the interests of all humanity and showing the farsightedness of a real statesman."

Alluding to Albanian and Chinese Communist criticism of the USSR's stand on Cuba, Tito said: "Unfortunately there are strange views about this policy . . . which are dangerous as they might lead certain circles to a false assessment of the motive for this wise and peace-loving action. I think that it is shortsightedness to consider these actions as a sign of weakness."

OAS Excludes Cuba

Castro Defies Threat. Cuban Premier Fidel Castro defied the Organization of American States Jan. 2 to carry out its threatened plans to exclude Cuba from its hemispheric activities.

Castro proclaimed his defiance of the other OAS member states before a Havana gathering of 500,000 followers at the 3d anniversary celebration of his revolutionary triumph over ex-dictator Fulgencio Batista. He declared: "We reiterate that we are Marxist-Leninists, and we do not repent it." He said Cuba had been forced to arm "to the teeth" to combat sabotage, anti-government conspiracies and the deliberate burning of cane fields. He called Cuba's military might "a defensive force" that was "infinitely superior" to that of Apr. 1961, when the Cuban exiles' invasion was defeated at the Bay of Pigs.

The revolutionary celebration was marked by a parade of Cuban military forces that included Soviet-built MiGs, multi-rocket launching units and anti-aircraft guns. The MiGs and other Soviet-built aircraft flew overhead. The military display was reviewed by Castro, his brother, Armed Forces Min. Raúl Castro, Pres. Osvaldo Dorticós Torrado and about 400 representatives of Communist countries and Latin-American leftist groups.

Punta Del Este Action. The foreign ministers of the Organization of American States voted Jan. 30-31 to exclude Cuba "from participation in the inter-American system." They did so on the grounds that Cuba's self-proclaimed adherence to Marxism-Leninism was "incompatible with the principles and objectives of the inter-American system" and that Cuba's "alignment . . . with the Communist bloc breaks the unity and solidarity of the hemisphere."

The ministers' action, designed to exclude Cuba from OAS activities but not to deprive Cuba of OAS membership, was taken at the final sessions of an OAS foreign ministers' conference begun Jan. 22 in Punta del Este, Uruguay. The U.S. had fought for stronger action against Cuba and had accepted compromises only in the hope of having the resolution adopted with near unanimity.

Cuba's ouster from inter-American affairs was specified in 3 of 9 resolutions approved by the ministers Jan. 30 and formally adopted Jan. 31. One resolution barred Cuba from the inter-American system. 2 others excluded Cuba from Inter-American Defense Board membership and imposed an arms embargo against the Castro régime.

The resolution excluding Cuba from OAS participation was approved by a 14-1 vote (6 abstentions), a ⅔ majority. Voting for: U.S., Colombia, Venezuela, Peru, Paraguay, Uruguay, Haiti, Dominican Republic, Panama, Honduras, Costa Rica, Nicaragua, Guatemala and El Salvador. Voting against: Cuba. Abstaining: Argentina, Brazil, Chile, Ecuador and Mexico.

Prior to the Jan. 30 vote, Cuban Pres. Osvaldo Dorticós Torrado told the conference that exclusion of his government from OAS affairs would cause the OAS' end as a regional system. He warned that the OAS' action would turn it into a "political-military bloc" of the U.S. Dorticós added: "You may expel us, but you cannot extract us from the Americas. You may put us out of the OAS, but the United States will continue to have a revolutionary and Socialist Cuba 90 miles from its shores."

U.S. State Secy. Dean Rusk described the approval of the anti-Cuba resolution as "a vivid demonstration of the democratic processes of a vig-

orous community of nations." In a speech at the final plenary session Jan. 31, Rusk said: "The friends of freedom must stand together . . . but it does not mean that we must be unanimous on all points. This is not a meeting of the Warsaw Pact." Eventually Cuba would be "released from its nightmare and returned to the family of the American states."

Colombian Foreign Min. José Joaquín Caicedo Castilla said the OAS action against the Castro régime, based on Article 6 of the Rio Treaty of Inter-American Reciprocal Assistance, was taken because Cuba was "a danger to the peace of the Americas."

Brazilian Foreign Min. Francisco San Tiago Dantas said the OAS should accept the fact that Cuba was a Communist-bloc member but should try to create "conditions for competition with Cuba in which we can be sure of victory for . . . democratic principles."

Dorticós, who boycotted the final OAS session with his delegation, said at a news conference before leaving for Cuba Jan. 31: "The United States came [to the conference] with the offer of dollars for the Alliance for Progress and made it clear that the award of the United States dollars depended on the position adopted here on Cuba."

(Pres. Kennedy, in a statement made at the opening of his news conference in Washington Jan. 31, expressed "satisfaction" with the OAS foreign ministers' action against Cuba. "For the first time," Mr. Kennedy asserted, "the independent American states have declared with one voice that the concept of Marxist-Leninism is incompatible with the inter-American system and they have taken explicit steps to protect the hemisphere's ability to achieve progress with freedom.")

The OAS Council, meeting in Washington Feb. 14, completed action on the organization's exclusion of Cuba. It did so by accepting the resolutions approved at the meeting in Punta del Este. Cuba's 3-man delegation, headed by Carlos M. Lechuga, walked out of the meeting shortly before the exclusion action was taken. The delegation withdrew after Lechuga's unsuccessful efforts to gain the floor to protest the Punta del Este decisions.

Argentina Ends Ties. Argentina broke off diplomatic relations with Cuba Feb. 8. It was the 14th Western Hemisphere nation to do so.*

Foreign Min. Miguel Angel Carcano said that Argentina's action was based on the measures urged against Cuba by the OAS. The Cuban chargé d'affaires was given 48 hours to leave Argentina. Argentine Amb.-to-Cuba Julio Amodeo had come back to Buenos Aires earlier Feb. 8.

(The Latin American Free Trade Association, by a 7-0 vote [2 abstentions: Mexico and Brazil] Sept. 4, rejected Cuba's membership application. Voting against Cuba: Argentina, Chile, Colombia, Ecuador, Paraguay, Peru and Uruguay. The 7 nations argued that Cuba could contribute nothing to the group and that its application had been made for political reasons only. The trade group was meeting in Mexico City.)

UN Rejects Cuban Charges. The Castro government went before the UN in February and March to charge that the U.S. had instigated the OAS to exclude Cuba illegally and that it was planning fresh interference in Cuba's affairs, possibly a repetition of the unsuccessful Apr. 1961 invasion. The UN refused to act on any of the Cuban charges.

The UN General Assembly's Political Committee Feb. 5 heard Cuban delegate Mario Garcia-Inchaustegui charge that the U.S. had gone to the

OAS' Punta del Este meeting "with a bag of gold in one hand and a bloody dagger in the other" to eliminate the possibility of negotiations with Cuba. He asked that the committee forbid the U.S. to carry out the "direct and unilateral intervention" it planned in Cuba with U.S. armed forces. The debate was concluded Feb. 15 when the committee voted by 50-11 (Cuba and Soviet bloc in favor, 39 abstentions) to reject a Czech-Rumanian resolution asking that the Assembly call on the U.S. to halt "interference" in Cuba's affairs and to desist from threats to Cuba's sovereignty or territory.

The Security Council was asked by Cuba Feb. 27 to place on its agenda charges that the U.S. had induced the OAS to exclude Cuba illegally. Ignoring Cuba's claim that the OAS action violated the UN Charter, the Council denied the request the ⅔ majority vote required for adoption of an agenda item. Cuba later succeeded in having the Council take up its complaint in another form, but its resolutions asking that the World Court rule on the OAS action and that the exclusion be set aside as illegal were rejected Mar. 23.

U.S. Tightens Embargo

Trade Ban Nearly Total. Pres. Kennedy Feb. 3 ordered an almost total embargo on U.S.-Cuban trade. He said the new embargo regulations, to go into effect Feb. 7, were aimed at depriving Cuba of the dollar exchange it had used "to engage in acts of aggression, subversion or other activities endangering the security" of the U.S. and other American nations.

The embargo halted all U.S. imports of Cuban tobacco and of some

---

* These other Western Hemisphere nations had ended relations with Cuba: the U.S., Dominican Republic, Guatemala, Nicaragua, Peru, El Salvador, Honduras, Costa Rica, Venezuela, Colombia, Panama, Paraguay and Haiti. Uruguay had withdrawn its ambassador from Cuba.

industrial molasses and vegetables. It permitted, "on humanitarian grounds," the export to Cuba of "certain [U.S.] foodstuffs, medicines and medical supplies." An embargo on exports to Cuba of nearly all manufactured goods, ships and certain foodstuffs had existed since Oct. 20, 1960.

The President said he acted under the authority of Section 620 (A) of the Foreign Assistance Act of 1961 and in accordance with the resolutions of the recent OAS foreign ministers' conference. Mr. Kennedy called the embargo a "necessary" step "to promote national and hemispheric security by . . . isolating Cuba and thereby" reducing "the threat posed by its alignment with the Communist powers."

Castro assailed the embargo Feb. 4 as "another economic aggression." In a speech at a Havana rally, Castro described as "hypocrisy" the U.S. decision to permit the continued shipment of U.S. food and medicine to Cuba. Cubans must triple their efforts in industry, agriculture and commerce to overcome the U.S. blockade, Castro said. He predicted "long years of sacrifice."

The U.S. embargo was tightened further Mar. 23 when the Treasury Department banned the import of all goods made from Cuban materials. Such imports had been permitted from non-Cuban countries under the Feb. 3 embargo proclamation. U.S. officials said the new order, to go into effect Mar. 24, was aimed largely at tobacco products. U.S. Sens. Spessard L. Holland and George A. Smathers (both D., Fla.) had sought the embargo extension to prevent the import from other countries of cigars made from Cuban wrapper tobacco.

(Canadian Prime Min. John G. Diefenbaker said Feb. 2 that Canada would consider what is "appropriate" for itself in deciding whether to continue to trade with Cuba. He said that during Jan.-Sept. 1961, U.S.-Cuban trade had been greater than Canadian-Cuban trade. During that period Canadian exports to Cuba had totaled $21½ million and Cuban exports to Canada $4,600,000 whereas U.S. exports to Cuba amounted to $13,300,000 and imports from Cuba $26,800,000, according to Diefenbaker. Diefenbaker denied that Castro promoted subversion with dollars derived from Canadian trade. "On the contrary, Cuban trade with Canada is absorbing dollars secured from other sources," he declared.)

U.S. Seeks Shipping Halt. The UPI reported Sept. 8 that the U.S. had urged its NATO allies to take these steps in helping to isolate Cuba: (a) Discourage their commercial ships from being chartered to the USSR to carry supplies to Cuba; (b) bar the use of their countries for the indirect shipment of U.S.-embargoed goods to Cuba; (c) refrain from exporting to Cuba goods that were of strategic, economic or military value; (d) prevent the extension of commercial credits to Cuba.

Turkey became the first NATO nation to agree publicly to U.S. requests to bar the use of its ships for carrying Soviet goods to Cuba. Turkish Foreign Min. Feridun Cemal Erkin said in New York Sept. 27 that his government, in compliance with a request earlier in the month by U.S. Amb.-to-Turkey Raymond A. Hare, had prevailed on private Turkish shipowners to withdraw voluntarily from the Cuban trade.

West Germany had adopted an ordinance requiring all its ships to obtain special permission to carry cargoes to Communist nations. The Bonn government said the regulation would "practically exclude" the shipment of strategic goods to Cuba.

Pres. Kennedy conferred in Washington Sept. 30 with Lord Home, British foreign secretary, on the problem of British vessels that continued to carry Soviet goods to Cuba. Despite their talks, David M. Robinson, president of the British Chamber of Shipping, declared after a meeting of the private shipowners' group in London Oct. 11 that British shipowners would continue to trade with Cuba.

Rusk met Norwegian Foreign Min. Halvard Lange in New York Sept. 24 to persuade Norway to bar the use of its ships for the sending of strategic goods to Cuba. The Norwegian Shipowners Association said in Oslo Oct. 1 that its members had been advised against sending their vessels to Cuba.

The Greek Shipowners Union in Athens was reported Oct. 10 to have recommended that its members "abstain totally . . . from concluding any transportation contracts to and from . . . Cuba" with Greek-flag ships.

(The U.S. Maritime Administration reported Sept. 29 that 20 Western nations' merchant ships had made 185 trips to Cuba during June-August; 61% of the runs were made by Greek, British, West German and Norwegian ships.)

U.S. Acts Vs. Shipping. Pres. Kennedy Oct. 4 set in motion a 4-point program to curb the use of U.S. or foreign ships for Cuban-Soviet bloc trade. The plan, to go into effect in 2 weeks, was embodied in an executive order.

The order: (1) Closed U.S. ports to all ships of any nation that permitted its vessels to carry military equipment to Cuba; (2) barred U.S. ports to any ship that, on a continuous run, brought non-military Communist cargoes to Cuba; (3) penalized foreign shipowners whose vessels engaged in the Cuban-Communist bloc trade by prohibiting them from carrying U.S. government cargoes on any of their ships; (4) forbade U.S.

flag ships or U.S.-owned ships from carrying goods to or from Cuba.

(An official of the British Chamber of Shipping Oct. 6 described the U.S. shipping curbs as "sanctions for carrying out perfectly legitimate trade." Canadian Foreign Affairs Min. Howard M. Green told Parliament Oct. 4 that his government would not prohibit Canadian ships from carrying goods between the USSR and Cuba because "Canada isn't directly involved in this whole development.")

Dorticós Scores Embargo. Cuban Pres. Dorticós, addressing the UN General Assembly Oct. 8, assailed the U.S. embargo as a "naval blockade" of Cuba. He urged the UN to condemn "such aggressive acts."

Dorticós said: "I ask" whether U.S. "pressure for blockade, the carrying out of unilateral acts to force this blockade . . . in a time of peace . . . is or is not

DORTICÓS At UN Assembly. (Wide World)

an act of war"; these "unilateral decisions" "flout and violate the principles" of the UN Charter; Cuba had no intentions of "carry[ing] out any activities extending the ideologies of its revolutionary processes to any other countries or parts of this hemisphere"; the "real hemispheric problem" was not Cuba but "hunger, famine, lack of culture, illiteracy" and the U.S.' "lack of respect for the sovereignty of other states."

Dorticós' speech was interrupted 4 times by anti-Castro demonstrators in the audience who shouted epithets at him. UN guards ejected 20 demonstrators from the Assembly hall.

Soviet Arms Shipments

U.S. officials reported in Washington Aug. 20 that an estimated 3,000 to 5,000 Soviet technicians had arrived in Cuba together with quantities of what appeared to be heavy military equipment. The Soviet personnel and equipment were reported to have been carried to the island in a fleet of 15 ships that arrived in Cuban ports between July 21 and 27.

Pres. Kennedy confirmed at his news conference Aug. 22 that "large quantities" of equipment and "an increased number of technicians" had come to Cuba from the Soviet bloc. But he said there was no conclusive evidence that this aid was exclusively military or that it involved Soviet troops.

According to the Aug. 25 N.Y. Times, Cuban exile sources in Miami had reported that the Russians and their supplies had been trucked to Pinar del Rio and Matanzas Provinces, in the western part of Cuba closest to the U.S., under Cuban army guard.

JFK Warns Cuba. Pres. Kennedy, citing his concern "over the recent moves of the Soviet Union to bolster the military power" of Cuba, declared Sept. 4 that the U.S. was determined to keep the Castro régime from "export[ing] its aggressive purposes by force or the threat of force." The President said it was "U.S. policy" to use "whatever means may be necessary" to bar Cuba "from taking action against any part of the Western Hemisphere."

The President issued his statement immediately after he had consulted with Congressional leaders on reports of Soviet military aid to Cuba.

Mr. Kennedy said: "Information has reached" the U.S. "in the last 4 days . . . which establishes without doubt that the Soviets have provided the Cuban government with a number of defensive missiles" and "radar and other electronic equipment . . . required for their operation." "We can also confirm the presence of several Soviet-made motor torpedo boats carrying ship-to-ship guided missiles" with a 15-mile range. "There is no evidence of any organized combat force in Cuba from any Soviet bloc country; of military bases provided for Russia; of a violation of the 1934 [Guantánamo] treaty; of the presence of ground missiles; or of other significant offensive capability either in Cuban hands or under Soviet . . . guidance." The U.S. "in conjunction with other Hemisphere countries will make sure that, while increased Cuban armaments will be a heavy burden to the unhappy people of Cuba themselves, they will be nothing more."

Pres. Kennedy, rejecting Congressional demands for an invasion of Cuba, had said at his news conference Aug. 29 that "I think it would be a mistake to invade Cuba because an action like that, which could be very casually suggested, can lead to very serious consequences."

(Sen. George Smathers [D., Fla.] proposed Sept. 2 that Cuba be invaded by a U.S.-sponsored military organization made up of Western Hemisphere nations. Sen. Strom Thurmond [D., S.C.], also calling for invasion, declared that the longer the U.S. "waits to expel communism from Cuba, the more difficult will be the job." Sen. Kenneth B. Keating [R., NY.] warned that the USSR might build missile bases in Cuba that could "threaten not only the U.S. but other Latin American countries." Sen. Homer E. Capehart [R., Ind.], speaking at a GOP meeting in Rockville, Ind., had said Aug. 27 that most of the "3,000 to 5,000" Soviet personnel sent to Cuba were combat troops, not technicians, and that Pres. Kennedy knew it. Capehart called for a U.S. invasion of Cuba before "the hundreds of Russian troops [in Cuba] grow into hundreds of thousands.")

Soviet Statement on Arms. The Soviet government warned Sept. 11 that a U.S. attack on Cuba or on Soviet ships carrying supplies to Cuba would mean nuclear war. The USSR charged that a Presidential request for stand-by authority to order 150,-000 military reservists to active duty was a "screen for aggressive plans" that would "inevitably lead to aggravating the international atmosphere." The Soviet warning declared that arms and military technicians had been sent to Cuba, but "exclusively for defensive purposes."

The statement said in part: U.S. "imperialists have been alarmed by the failure of the . . . [U.S.'] economic blockade . . . to strangle the Cuban people"; the U.S. was "so much frightened" of the Soviet revolution "that it seems to you some hordes are supposedly moving to Cuba when potatoes or oil, tractors . . . and other farming and industrial machinery" were shipped to Cuba; "we can say to these people that these are our ships, and what we carry in them is no business of theirs"; the USSR was shipping to Cuba on Cuban request "a certain amount of armaments" "designed exclusively for defensive purposes"; the USSR also was sending Cuba "military specialists and technicians" "because up-to-date weapons now call for high skill and much knowledge"; "the number of Soviet military specialists sent to Cuba is in no way to be compared to the number of workers in agriculture and industry sent there"; "If the aggressor makes an attack on one state or another and this state asks for assistance, the Soviet Union has the possibility to render assistance to any peace-loving state and not only to Cuba"; "let no one doubt that the Soviet Union will

Men described as Soviet technicians and military personnel sing as they disembark in Havana during first week of September. (Photo from *Wide World*, obtained by *Bohemia Libre*, magazine formerly edited in Havana but currently published by Cuban exiles in New York).

render such assistance."

According to the Soviet statement, there was "no need for the Soviet Union to shift its weapons for the repulsion of aggression, . . . to any other country, for instance Cuba," because "our nuclear weapons are so powerful" and the USSR "has so powerful rockets" to deliver them "that there is no need to search for sites for them" outside the USSR.

Kennedy Reply. Pres. Kennedy, in a preliminary statement read at his news conference Sept. 13, declared that if the U.S. "ever should find it necessary to take military action against communism in Cuba, all of Premier Fidel Castro's Communist-supplied weapons and technicians will not change the result."

Mr. Kennedy reiterated that the U.S. would "do whatever must be done" "if at any time . . . Cuba were to endanger or interfere with our security in any way, including our base at Guantánamo, our passage to the Panama Canal, our missile and space activities in Cape Canaveral, or the lives of American citizens in this country," or by Cuban "aggressive" threats against "any nation in this hemisphere."

The President, however, again stressed that U.S. "unilateral military intervention" "cannot currently be either required or justified, and it is regrettable that loose talk about such action in this country might serve to give a thin color of legitimacy to the Communist pretense that such a threat exists."

Mr. Kennedy added a clear warning that the U.S. would consider it "a change in the nature of the [Cuban] threat" warranting strong action if Cuba became a major Soviet military base or if Soviet offensive missiles appeared on the island.

(Administration sources estimated Sept. 20 that: 65-75 shiploads of Soviet military items and personnel had been delivered to Cuba during the previous 2 months; about 4,200 Soviet military technicians, instructors and construction workers were in Cuba; the build-up was mostly related to anti-aircraft missile installations; 12 installations for ground-to-air missiles with a range of up to 25 miles had been detected in Cuba; in eastern Cuba near the U.S.' Guantánamo naval base the Russians had helped build a base for anti-ship missiles with a 20-35-mile range; 8 patrol boats, each with 2 guided missiles with ranges of 11-17 miles, were in Cuban waters; Cuba had about 60 old-style MiGs and at least one new MiG-21 jet fighter.)

Congress Backs Force. The U.S. Congress Sept. 26 completed action on a joint resolution stating the U.S.' determination "to prevent by whatever means may be necessary, including the use of arms, the Marxist-Leninist régime in Cuba from extending by force or theat of force, its aggresive or subversive activities to any part of this hemisphere."

The resolution, signed by Pres. Kennedy Oct 4, was in response to the Soviet-supplied arms build-up in Cuba. It declared that the U.S. was determined (1) "to prevent in Cuba the creation or use of an externally supported military capability" endangering U.S. security and (2) to

"work with the Organization of American States and with freedom-loving Cubans to support the aspirations of the Cuban people for self-determination."

U.S., Latins Study Action. Latin American foreign ministers and State Secy. Rusk held an informal meeting in Washington Oct. 2-3 to discuss measures to combat "the Sino-Soviet intervention in Cuba," which the meeting described as "an attempt to convert the island into an armed base for Communist penetration of the Americas."

Pres. Kennedy, in an address to the Latin American foreign ministers at a White House luncheon Oct. 2, said: "What we have to do is prevent the expansion from Cuba itself of the Communist doctrine using Cuba as a base and using Cuba as a spearhead to attack the security of all. Therefore, we believe it is in the interest of us all that we . . . consider what actions can be most carefully taken to contain the expansion of communism from . . . Cuba and to take those steps which will finally provide for the freedom of the Cuban people."

A communiqué issued at the end of the meeting Oct. 3 said:

"A military intervention of Communist powers in Cuba cannot be justified as a situation analagous to the defensive measures adopted in other parts of the free world in order to face Soviet imperialism"; the Western Hemisphere nations should "intensify measures to prevent agents and groups of international communism from carrying on" subversion; the Western Hemisphere nations should "intensify individual and collective surveillance of the delivery of arms and implements of war and all other items of strategic importance" to Cuba "in order to prevent" their use "for offensive purposes against the hemisphere."

The exchange of U.S. and Russian statements on the arming of Cuba and the Washington meeting of Latin and U.S. leaders were the last significant developments to precede the U.S.' announcement that it had discovered offensive Soviet missile sites under construction in Cuba.

Exiles & Prisoners

Cuban Raids Continue. The anti-Castro Cuban exile groups based in the southeastern U.S. and in Puerto Rico continued small-scale raids on Cuban coastal targets during 1962. No invasion attempt was made nor were any of the raids known to have inflicted serious damage to Cuban defenses.

Major raids reported by the exiles:

▶ The Havana suburb of Miramar was shelled Aug. 24 from 2 small motor boats manned by 23 members of the Miami-Based Student Revolutionary Directorate. Cuban planes, hampered by poor weather, made an unsuccessful attempt to intercept the launches, which made the attack from about ½ mile offshore. The boats put into port safely Aug. 25 at Marathon, Fla., where they were impounded by the U.S. Coast Guard pending an investigation to determine whether the exiles had violated U.S. neutrality laws. A statement issued by the Student Directorate Aug. 25 said that the 2 vessels had "bombarded the Chaplin Theatre" in Miramar with 30-60 shots from 20-mm. cannon "while a meeting of Russians and Czechoslovaks recently arrived in Cuba was being held." Several hotels and other buildings were hit. (The U.S. State Department said Aug. 25 that the exiles' attack "was a spur-of-the-moment raid" in which the U.S. government had not been involved and of which it had no previous knowledge.)

▶ Havana radio reported Sept. 11 that a "pirate vessel" (described as a 40-foot launch) operated by "criminals armed and paid by" the U.S. had attacked a Cuban dock boat and a British freighter Sept. 10 at the north coast port of Caibarién, 210 miles southeast of Havana. Cuban reports said that the British ship, the 7,043-ton *Newlane,* had been fired on 13 times as it was loading 31,000 sacks of sugar. The Cuban ship, the San Pascual, was hit 18 times. A Cuban exile group, identifying itself as "Alpha 66, Puerto Rico," issued in Miami Sept. 11 a communiqué saying it was responsible for the Caibarién attack. The communiqué said: "Cuban patriots," participating in a 50-minute naval engagement, "machine-gunned intensely" the *Newlane* and then "boarded, dynamited and machine-gunned the *San Pascual*"; the raiders also had attacked a 2d Cuban ship.

▶ Alpha 66 announced in New York Oct. 10 that a small group of its commandos had attacked a military camp at the Cuban port of Isabela de Sagua (about 150 miles east of Havana) Oct. 8 and had killed about 20 persons, including some Russians. Alpha 66 said the invaders had dynamited a railway switchyard, an arsenal, a commissary and 4 other strategic points and then had fled with captured weapons and ammunition. Some of the attackers were said to have been wounded.

▶ A vessel manned by 6 members of an unidentified Cuban exile group sank a Cuban patrol boat the night of Oct. 12-13 after attacking it with gunfire off the northern coast of Matanzas Province. 2 of the patrol boat's 4 crewmen were wounded and picked up by the exiles' vessel, which had embarked on the raid from the Bahamas. The 2 wounded Cuban prisoners were taken to the small British island of Cay Sal off northern Cuba, where they were picked up by a U.S.

Coast Guard helicopter Oct. 14 and flown to a Miami hospital.

The formation of a Cuban Revolutionary Junta to overthrow the Castro régime was agreed on at a meeting of Cuban exile leaders in Río Cañas, P.R. Sept. 16 and announced in New York Sept. 20. Junta leaders: ex-National Railways Min. Raúl Chibás; ex-Public Works Min. Manuel Ray; José Gómez Rodríguez and Rogelio Cisneros, coordinators of anti-Castro forces in Cuba; José R. San Martin.

Pres. Kennedy was reported Oct. 25 to have received a letter on behalf of 8 Cuban underground organizations declaring they would not accept any government "imposed" by the U.S. if and when Castro was overthrown. The telegram was sent from Miami by Col. José Piñeiro Paez, who said he represented the underground groups in the U.S. Piñeiro said the telegram advised Mr. Kennedy that the underground groups had selected Dr. Mario Garcia Kohly June 20 as "provisional president of the Cuban Government in arms in exile."

PWs Ransomed, Freed. 1,113 prisoners captured in the 1961 Bay of Pigs invasion of Cuba by anti-Castro exiles were released in exchange for $53 million worth of baby food and medicine under an agreement reached Dec. 21 in Havana by Premier Castro and James B. Donovan, New York attorney representing the Cuban Families Committee for the Liberation of Prisoners of War. The U.S. government had given strong behind-the-scenes support to the talks. The prisoners, members of Brigade 2506, were flown to Miami Dec. 23-24 in an airlift of 4 Pan American airliners.

The final obstacle to an agreement had been overcome when Donovan acceded to Castro's demand for payment of $2,925,000 in cash for 60 wounded Cuban prisoners released Apr. 14. The Cuban Families Committee at the time had promised to pay the ransom but later had been unable to do so. The money was raised Dec. 23 through the personal efforts of Atty. Gen. Robert F. Kennedy, Gen. Lucius Clay, ex-Treasury Secy. Robert B. Anderson and ex-Treasury General Counsel Robert Knight, who contacted private donors. Several other prisoners also had been freed after payment of their ransoms individually by friends and families.

As a "bonus," Castro agreed to let 923 relatives of the released prisoners leave Cuba to join their families in Miami. The relatives arrived in Miami Dec. 27 aboard the U.S. freighter S.S. *African Pilot,* which had brought the first cargo of food and medicine (value: $11 million) to Cuba Dec. 23. The shipment of all the supplies, organized by the American Red Cross, was expected to take several months. The food and drugs had been contributed by more than 100 U.S. manufacturers.

Donovan had first started negotiations with Castro in August, but the talks had collapsed Oct. 28 when the Cuban missile crisis erupted. Donovan had resumed his talks early in December after the crisis subsided. The Cuban Families Committee members who served with Donovan during his negotiations with Castro were Mrs. Berza Barreto, Mrs. Virginia Betancourt and Alvero Sánchez Jr.

A ransom of $62 million originally had been set for the prisoners; this amount was fixed in April after the 1,179 prisoners then under detention had been tried and convicted on treason charges Mar. 29-Apr. 3 by a Havana military tribunal. The prisoners, all sentenced Apr. 7 to 30 years' imprisonment and deprivation of citizenship, had been "priced" individually at $25,000 to $500,000. The latter figure was set for 3 of the leaders of the 1961 attack: Manuel Artime,

José Pérez San Roman, and Eneido Oliva.

JFK Greets Captives. Pres. Kennedy reviewed the 1,113 newly-freed prisoners at a rally of 40,000 Cuban exiles Dec. 29 at the Orange Bowl Stadium in Miami. The exiles presented to Mr. Kennedy the battle flag of their brigade. The President expressed "great appreciation for making the U.S. the custodian of this flag." "I can assure you that this flag will be returned to this brigade in a free Havana," the President declared.

Paying tribute to the exiles' courage and cause, Mr. Kennedy said: "I can assure you that it is the strongest wish of the people of this country, as well as the people of this hemisphere, that Cuba shall one day be free again, and when it is, this brigade will deserve to march at the head of the free column."

The President's speech evoked cries of *"Guerra! Guerra!"* (War! War!) and *"Libertad"* (Freedom).

Mrs. Kennedy, whose son, John, 2, had seen some of the brigade officers when the President had received them Dec. 27 at his Palm Beach home, delivered a brief speech in Spanish. She said: "I feel proud that my son has known the officers. . . . It is my wish . . . that some day he may be a man at least half as brave as the members of Brigade 2506."

Other Developments

3 Cubans Seized in U.S. 3 pro-Castro Cubans were arrested in New York Nov. 17 and indicted by a federal grand jury Nov. 21 on charges of conspiring to commit sabotage in the New York area.

FBI agents had arrested them after discovering a cache of arms and explosives in a Manhattan (N.Y. City) jewelry office owned by one of the ac-

Pres. and Mrs. Kennedy honor former Cuban invasion prisoners at ceremony in Miami Dec. 29. Among those present: Miami Mayor Robert King High (left) and the Rev. Ismael Lugo (right), chaplain of invasion brigade.   (Wide World photo)

cused. 2 attachés of the Cuban mission to the UN—José Gómez Abad, 21, and his wife, Elsa Montero de Gómez Abad, 20—were accused of passing weapons to the accused saboteurs. The U.S. demanded their recall, and the pair left for Cuba Nov. 19.

The 3 alleged saboteurs: Roberto Santiestebán Casanova, 27 (he had arrived in the U.S. Oct. 3 on a diplomatic passport as a member of the Cuban mission to the UN, but the Justice Department contended that he was not entitled to diplomatic immunity because his passport was still being processed at the time of his arrest); Marino Antonio Estebán del Carmen Sueiro y Cabrera, 22; José García Arellana, 42, owner of the jewelry office in which the arms were found.

At the arraignment Nov. 17 before Federal Commissioner Earle N. Bishopp, U.S. Atty. Vincent L. Broderick charged that the 3 defendants had planned to use the arms "for the purpose of injuring and destroying national defense materials, premises and utilities" in the New York area.

Venezuelan Oil Dynamited. Saboteurs, believed to be members of a pro-Castro group, dynamited 4 electric power stations of the U.S.-owned Creole Petroleum Corp. (a subsidiary of Standard Oil of N.J.) in the area of Lake Maracaibo, Venezuela Oct. 27. 2 of the saboteurs were captured and a 3d was reported to have been killed. A 2d group of saboteurs Nov. 2 dynamited 4 pipe lines of oil installations operated by U.S. companies at Puerto La Cruz, 260 miles east of Caracas. The saboteurs' action caused considerable damage to 3 oil pipelines operated by the Mobil Oil Co. and the Texas Oil Co.

Havana radio asserted Oct. 29 that the first dynamiting had been carried out by "the Army of Venezuelan Liberation" as its "first reply . . . to the military mobilization decreed by the puppet Rómulo Betancourt."

Ben Bella in U.S. & Cuba. Algerian Premier Ahmed Ben Bella flew to New York Oct. 6 on a visit to the UN, followed by meetings with Pres. Kennedy in Washington and Cuban Premier Fidel Castro in Havana.

Ben Bella flew to Washington Oct. 14 and conferred privately with Pres. Kennedy Oct. 15 at the White House. He then flew to Havana, where he was welcomed with a warm embrace Oct. 16 by Premier Castro.

Castro, in a welcoming speech, said Ben Bella's visit was "an act of courage" in the face of "Yankee imperialist hostility." Ben Bella replied that Algeria was grateful for Cuba's aid despite Cuba's "enormous difficulties." Every Algerian "knows, has lived and admired" the Castro revolution "in all its phases," Ben Bella said. The Cuban government announced Oct. 17 that Ben Bella had indorsed Castro's demand that the U.S. vacate its naval base at Guantánamo.

Ben Bella flew back to New York Oct. 17 and returned to Algiers Oct. 20.

(The U.S. State Department Oct. 23 reportedly ordered the Agency for International Development to suspend indefinitely discussions with Algerian officials on possible U.S. economic aid to Algeria. The *N.Y. Times* said Oct. 26 that the suspension apparently was due to Pres. Kennedy's displeasure with Ben Bella's visit to Cuba. Foreign Min. Khemisti told the National Assembly Nov. 26 that U.S. aid talks had been suspended because of the Cuban blockade crisis and would be resumed "very shortly." Khemisti said that the Kennedy-Ben Bella talks had been "extremely positive" and that Ben Bella's Cuban trip had been just a "fraternal" gesture natural to a nation with Algeria's policy of neutrality in the cold war.)

Politics & Government. The formation of a 25-member directorate to lead the Cuban revolution was announced Mar. 9 by the government of Premier Fidel Castro Ruz.

The directorate, which included Castro, his brother, Armed Forces Min. Raúl Castro Ruz, and Industry Min. Ernesto (Ché) Guevara was to become the leading organ of Cuba's Integrated Revolutionary Organizations (ORI), the unitary proletarian political party that Castro had promised to create to rule Cuba and to lead its revolution. The ORI was described officially in the announcement as Cuba's "Marxist-Leninist Party." 10 of the 25 members of its ruling directorate were known Communists.

Other directorate members (Communists identified by the letter 'C'): Pres. Osvaldo Dorticós Torrado; CP leader Blas Roca; revolutionary militia chief Emilio Aragones; Labor Min. Augusto R. Martínez Sánchez; National Institute of Agrarian Reform Pres. Carlos Rafael Rodríguez (C); Cuban Amb.-to-USSR Faure Chomon; Interior Min. Ramiro Valdes; Institute of Agrarian Reform official Severo Aguirre (C); Cesar Escalante (C); ex-Sen. Joaquín Ordoqui (C); Flavio Bravo (C); Cuban Workers Federation head Lazaro Peña (C); Manuel Luzardo (C); Ramón Calcines (C); Maj. Juan Almedia, army chief; Education Min. Armando Hart Davalos; Haydee de Santamaria Hart, Armando Hart's wife; Maj. Sergio del Valle, air force head; Public Works Min. Osmani Cienfuegos; Communications Min. Raúl Curbelo; Guillermo García; Anibal Escalante (C).

It was announced Mar. 23 that Castro had been named first secretary and his brother, Raúl, had been named 2d secretary of the ORI directorate. Anibal Escalante, a prominent Communist member of the directorate, was made its organizing secretary.

(But Escalante was denounced by Castro Mar. 26 for having brought "chaos to all the country" by using his position "to pursue personal ends." Escalante, widely regarded as the theoretician of the Cuban revolution, was dismissed from the ORI directorate Mar. 27. He was reported to have left Havana for Prague Mar. 26.)

Raúl Castro was appointed Mar. 25 to the newly-created post of vice premier in the Cuban government.

In other cabinet changes announced Mar. 5 Manuel Luzardo replaced Maximo Berman as interior trade minister; Celia Sánchez replaced Luis Buch Rodriguez as secretary to the Presidency and the Council of Ministers (Buch had been named Supreme Court head); Amb.-to-USSR Faure Chomon was named communications minister, replacing Capt. Raúl Curbelo, who was appointed air force chief. Curbelo filled the vacancy created by the appointment of Cmdr. Sergio del Valle to the ORI directorate. (Lt. Cmdr. Rolando Díaz Aztarain, finance minister, was put in command of the navy June 6. Díaz Aztarain was replaced as finance minister by Luis Alvarez Rom.)

Economy. A record budget of 1,853,500,000 pesos for 1962 was approved by the Council of Ministers and announced Jan. 6. (Peso exchange rate: 20-25 to the dollar.) Budget items (figures for defense were not made public): education—237 million pesos; health—89 million; science & culture—32 million; communications —48 million; social security—175 million; administrative services—194 million; recreation & sports—9 million; agriculture, livestock & fisheries—111 million; industrial expansion—208 million; transportation—88 million; social welfare—551 million; basic community services—228 million; "national economy & construction of socialism" —696 million. (The sum of the itemized figures was higher than the announced 1,853,500,000-peso total budget.)

Carlos Rafael Rodríguez, 48, editor of the Communist newspaper *Hoy*, an authority on Marxist economic theory (economics professor at Havana University), replaced Premier Castro Feb. 14 as president of the National Institute of Agrarian Reform.

It was reported Feb. 25 that the Integrated Revolutionary Organizations (ORI) had appointed a 3-man committee to direct Cuba's economy. The committee's principal function: to cope with a supply crisis, particularly in food. Its members: Pres. Dorticós; Industry Min. Guevara; Pres. Rodríguez of the National Agrarian Reform Institute.

Premier Castro announced in a TV broadcast Mar. 12 that these basic foods and other essentials would be added to the list of items already rationed on a nationwide basis: rice and beans; lard and cooking oils (rationed in most areas since 1961); vegetables, milk, fish, chicken and beef in the Havana area; soap, toothpaste and detergents. Castro said restaurant supplies would be reduced by as much as 50% to keep people with money from buying all the food they wanted. Castro attributed the scarcities to the U.S.' "brutal economic blockade." (The Cuban Health Ministry had ordered the rationing of drugs and medicines Feb. 11. Gasoline had been rationed since Jan. 28.)

The National Agrarian Reform Institute announced Aug. 18 that cooperative farms would be replaced by government-operated "people's farms" because the cooperatives had been mismanaged. (The cooperative farm system had been established by the Agrarian Reform Law of May 1960.) Institute Pres. Rodríguez, speaking at a national meeting of cooperative workers in Havana, said the decision to replace cooperatives had been made after 90% of the 118,000 cooperative farm workers had voted in favor of the move. (The decision to replace cooperatives with "people's farms" was the subject of a speech by Premier Castro in Havana Aug. 19. Speaking at the closing session of the Congress of Sugar Cane Cooperatives, Castro declared that the distribution of land to the peasants "would have ruined the revolution." He estimated that there were 600 cooperatives of all types and 300 "people's farms" in Cuba. Castro warned of continued food shortages despite the farm system change.)

Increased government control of labor was announced in a decree signed Aug. 29 by Labor Min. Martínez Sánchez. Major points of the decree: (a) "All pay raises . . . without the knowledge and approval of the Labor Ministry will be null and void"; (b) workers may be transferred from one place of employment to another, either as a penalty for absenteeism or to help meet production needs elsewhere.

The government Dec. 5 nationalized privately owned retailing and wholesale stores and businesses dealing in clothing, shoes and hardware.

Soviet Relations. The AP reported Jan. 10 that the Soviet and Cuban governments had signed a trade pact providing for the exchange of $700 million worth of goods during 1962. This was an increase of $150 million over the 2 nations' trade in 1961. According to the report, Cuba was to send the USSR sugar, alcohol, nickel, rum and tobacco; the USSR was to ship oil and petroleum products, metals, fertilizer and industrial equipment.

(Tass reported Aug. 28 that Russia had sent Cuba goods valued at 284 million rubles in 1961 and had received from Cuba 281 million rubles worth of goods. [One ruble=$1.11.])

A program of increased Soviet military and industrial assistance to Cuba was announced in a joint Cuban-USSR communiqué issued in Moscow Sept. 2. The bilateral agreements were reached following talks that had started in Moscow Aug. 27 and whose major participants were Cuban Industry Min. Guevara, Emilio Aragones Navarro, head of the Cuban militia, Premier Khrushchev and other Soviet leaders. The amount and type of Soviet military aid to Cuba was not specified. The communiqué only said that "in view of the threats of aggressive imperialist quarters with regard to Cuba," the USSR had agreed to Cuba's "request for help by delivering armaments and sending technical specialists for training Cuban servicemen." The communiqué disclosed that under an agreement signed Sept. 1 the USSR would help Cuba build a metallurgical factory and expand the output of 3 existing metallurgical factories from 110,000 tons a year to 350,000.

Castro announced Sept. 25 that the USSR had agreed to help build a port in Havana Bay as headquarters for a joint Cuban-Soviet fishing fleet.

Executions & Plots. The execution by firing squad of anti-Castro guerrilla leader Margarito Lanza Flores was announced by the government

Jan. 4. Lanza was identified as the leader of a band accused of killing 3 militiamen.

3 Cubans were executed Jan. 27 for allegedly sabotaging urban transportation with incendiary bombs supplied by the U.S. Those executed: Rolando Rocha Menendez, René López Menendez, Manuel López Dominguez.

Braulio Amador Quesada, 47, leader of the underground United Catholic Movement, was executed Jan. 30 on counter-revolutionary charges. Amador was identified as one of the 1961 Cuban rebel invaders.

Cuban security officials announced Nov. 13 that the leader of an armed band of U.S.-trained saboteurs had been arrested Nov. 2 near Vinales.

The arrest, according to the officials, thwarted an attempt by the saboteurs to blow up the Matahambre copper mines at the western tip of Cuba. The leader of the band was identified as Miguel Angel Orozco Crespo, a lieutenant in Batista's army.

Castro Excommunicated. Msgr. Dino Staffa, secretary of the Vatican's Sacred Congregation of Seminaries & Universities, disclosed Jan. 3 that Premier Castro and other Cuban officials had been automatically excommunicated for "impeding and imprisoning" Roman Catholic bishops in Cuba. (Havana Bishop Eduardo Boza Masvidal and 135 Roman Catholic priests had been arrested and deported from Cuba Sept. 17, 1961.)

Press reports charged that Soviet forces in Cuba were engaged in extensive military base construction. The work was reported to be concentrated (A) between Havana and the San Antonio de Los Baños airbase and (B) in the Sierra Maestra area.   (Wide World map)

In the wake of 1962's crisis over Soviet missiles in Cuba, the U.S. became embroiled in controversy in 1963 over charges that the USSR was still arming the island. The U.S. and OAS tightened economic restrictions on Cuba in the face of accusations that Cuba was the focus of Communist subversion against Latin America. Cuban Premier Castro visited Russia in 1963 and won new assurances of military backing should the U.S. invade Cuba. Cuban exiles raided the island, and underground groups in Cuba attacked repeatedly during 1963.

Missile Crisis Aftermath

The Soviet-U.S. confrontation in 1962 over the construction of Russian missile bases in Cuba had brought the world nearly to the edge of conflict but had been settled in a series of diplomatic maneuvers directed by the late Pres. John F. Kennedy and Soviet Premier Nikita S. Khrushchev.

The U.S., which had blockaded Cuba and had appeared ready to strike at the Soviet missile sites, relented only after the Soviet Union, in an exchange of cabled messages and in negotiations carried out at the United Nations, had agreed to dismantle the bases and to withdraw the rockets and medium-range jet bombers it had sent to Cuba. Soviet ships sailed from Cuba with dismantled missiles and jet bombers displayed on their decks, while U.S. naval vessels and aircraft steamed alongside and flew low overhead to count the weapons and verify their removal from the island.

Cuba itself was subjected to intensive aerial surveillance—much of it by

high-altitude U-2 planes—to assure that the missile launching pads and installations had been dismantled.

Despite evidence that the USSR had carried out its pledge to remove the missile threat from the Caribbean, and despite a public declaration by the 2 powers that they considered the matter closed, there were recurrent reports that the USSR had hidden aggressive arms in Cuba and that the island remained a potential base for action against the U.S. and other Western Hemisphere nations. These reports, denied by the Kennedy Administration, were accepted as authentic by many Congressional leaders opposed to the Administration's Cuban policies, and became the source of a new and continuing U.S. political controversy in 1963.

Crisis Ended Officially. The U.S. and the USSR announced Jan. 7 that they regarded the Cuban missile crisis as closed and that they had ended their UN negotiations on it.

The announcement was made in a joint letter delivered to UN Secy.

Gen. U Thant by the 2 chief negotiators in the talks: U.S. Amb.-to-UN Adlai E. Stevenson and Soviet Deputy Foreign Min. Vasily V. Kuznetsov. Although the letter acknowledged that certain aspects of the dispute remained unsettled, it asserted that sufficient progress had been made to remove the matter from the agenda of the UN Security Council.

One of the unsettled facets referred to by the letter apparently was the U.S.' refusal to issue a formal pledge against invading Cuba until it was granted some form of on-site inspection to verify the removal of Soviet missiles and other aggressive weapons from the island.

The U.S. also informed the Council of the Organization of American States Jan. 7 that negotiations on the missile crisis had been brought to a successful conclusion.

Castro Vs. Terms of Accord. Cuba's opposition to the Soviet-U.S. agreement to remove missiles from Cuba was expressed by Premier Fidel Castro Jan. 2 in a speech at a Havana rally marking the 4th anniversary of the Cuban revolution's overthrow of the Batista dictatorship.

Referring pointedly to Cuba's intention to exercise its sovereignty and independence despite the Soviet-U.S. accord, Castro said: "The Soviet government has reached certain accords with the American government. But this does not mean that we have renounced the right to have the weapons we deem convenient and to take steps in international policy that we deem convenient as a sovereign country."

Castro spoke after reviewing a 2-hour parade of Cuba's armed forces that included Soviet-built MiG jet fighter planes, helicopters, tanks, ground-to-air rockets and anti-aircraft rockets. The MiGs, apparently flown by Cuban pilots, included the MiG-21 supersonic interceptor. In the review-

ing stand with Castro were about 400 foreign Communist leaders and delegates.

Rusk on No-Invasion Pledge. State Secy. Dean Rusk appeared before the Senate Foreign Relations Committee in Washington Jan. 11 to clarify the U.S. position on the pledge it had given not to invade Cuba if the USSR removed its aggressive weapons from the island. Pres. Kennedy had given the pledge to Premier Khrushchev in an exchange of letters Oct. 27-28, 1962, at the height of the missile crisis.

After the closed hearing, Committee Chrmn. J. W. Fulbright (D., Ark.) quoted Rusk as saying that Mr. Kennedy's pledge had been "strictly a contingent commitment" conditioned on the acceptance by the USSR and Cuba of on-site inspection to make certain that Soviet missiles were removed from Cuba. Rusk reportedly had said that since Cuba barred on-site inspection, the U.S. had not committed itself to a no-invasion pledge despite Pres. Kennedy's Nov. 20, 1962 statement that the U.S. had no intention of invading Cuba.

The Senators also queried Rusk as to why no definite intelligence had been obtained on the presence of Soviet missiles in Cuba until the U-2 flight had discovered them Oct. 14, 1962. (At least one high U.S. official had reportedly suggested in mid-Aug. 1962 that reconnaissance flights be conducted over Cuba on the basis of reports from Cuba itself that missiles had arrived there.) Rusk replied that intelligence reports received by the U.S. in Aug.-Sept. 1962 had not been definite enough "to base policy on."*

* Defense Secy. Robert S. McNamara was questioned by a House Appropriations subcommittee Feb. 7 on the question of why it had taken until Oct. 14, 1962 to discover missile sites in the San Cristobal area, which apparently had been under suspicion since the previous Sept. 21. In testimony released Mar. 29, McNamara asserted that: (a) the first ships carrying the Soviet missiles did not arrive in Cuba before Sept. 8; (b) construction on the sites did not

Soviet Build-Up in Cuba Disputed by U.S. Leaders

Island Reported Red Base. Reports appearing in the U.S. press in March warned repeatedly that the USSR had resumed shipments of arms to Cuba and that numbers of Soviet military men were helping the Castro régime weld these arms into a highly sophisticated ground and air defense system. These reports, stemming largely from Cuban exile sources, persisted despite emphatic denials of their truth by Pres. Kennedy and other Administration officials.

One of the most comprehensive of the reports, published Jan. 23 by the Western edition of the *N.Y. Times*, said that "large shipments of weapons" had arrived in Cuba from Russia. It gave the following account of military developments on the island:

▶ Soviet encampments were being relocated and strengthened. Soviet experts were directing Cubans in the construction of underground depots, hangars and runways. The construction apparently was concentrated between Havana and the San Antonio de Los Baños airbase in Camagüey Province and in the Sierra Maestra region of Oriente Province.

▶ The Soviet weapons brought into Cuba were of a "defensive" nature, as far as could be determined.

▶ The number of Russians in Cuba during the recent missile crisis was estimated at 15,200. About 3,300 (presumably attached to the dismantled missile sites) had been withdrawn. But replacements brought the USSR's current force in Cuba to more than 12,000 men.

▶ An estimated 6,000 Soviet troops were manning 24 ground-to-air missile sites in all parts of Cuba.

▶ Soviet specialists were handling tactical battlefield rockets (known to U.S. military men as "frogs") with a nuclear capacity and an over-the-horizon range of 15-25 miles. U.S. experts had no positive evidence that nuclear tips for those weapons were actually in Cuba.

Congressional Charges. The reports prompted the Senate Foreign Relations Subcommittee on Inter-American Affairs, headed by Sen. Wayne L. Morse (D., Ore.), to hold a one-day closed hearing in Washington Jan. 25. State Secy. Rusk and CIA Dir. John A. McCone appeared at the meeting.

Among statements of committee members after the meeting: Sen. George D. Aiken (R., Vt.)—"Russia has built an enormously strong political and military base in Cuba—much stronger than it was 6 months ago—and it is still continuing." Sen. Stuart Symington (D., Mo.)—There had been "an additional build-up" in Cuba in the past several months. "We ought to begin to worry about what is in Cuba. There is a big Soviet military complex in Cuba."

Chrmn. John Stennis (D., Miss.) of the Senate Preparedness Subcommittee attended the meeting and said his group would hold hearings on the reported Soviet build-up in Cuba. Stennis said that reports indicated that the build-up "is rapidly converting Castro's forces into the 2d most powerful military force in the hemisphere."

Sen. Kenneth B. Keating (R., N.Y.) asserted in the Senate Jan. 31 that "a large [Soviet] vessel" had arrived in Cuba Jan. 25 and that, "under the highest security conditions, it unloaded a cargo of armaments." Keating said the ship and another Soviet vessel had followed a route to Cuba "through areas where the U.S. is least

start before Sept. 15-20; (c) none of the sites had attained operational status before their discovery Oct. 14. Reporting on the U-2 missions over San Cristobal in Sept.-Oct. 1962, McNamara said a Sept. 5 flight over San Cristobal had "showed absolutely no activity." He conceded that reports were received Sept. 20-21 of suspicious activity in San Cristobal, but that no further flights had been conducted over the area until Oct. 14 because attention had shifted to the other end of Cuba.

able to maintain adequate surveillance of ships' contents."

Keating also reported that "there is absolutely confirmed and undeniable evidence that the Soviets are maintaining the medium-range missile sites they had previously constructed in Cuba. This gives rise to the very real possibility that Russia hopes to return the heavy missiles to the island—or, even more ominous, that they may have left missiles on the island and need only to wheel them out of caves."

Sen. Strom Thurmond (D., S.C.), declaring that Cuba had become "a formidable Soviet strategic military base," said Feb. 1 that "reliable sources" estimated Soviet strength in Cuba at: 30,000-40,000 troops with supporting air and naval units; 600 tanks, 200 armored personnel carriers, more than 2,000 artillery pieces, more than 1,000 mortars; slightly less than 1,000 anti-aircraft missiles; 4 battalions of nuclear-tipped 300-mile-range "frog" missiles; more than 300 planes and helicopters.

Rep. Donald C. Bruce (R., Ind.) told the House Feb. 4: There were 40-50 offensive missiles in Cuba, "and the highest officials of the U.S. government know it"; "highly placed officials in Cuba from several major Western powers friendly to the U.S. counted the unloading from Soviet vessels in Cuba of 82 to 88 Soviet missiles—not 42."

JFK Denies Direct Threat. Pres. Kennedy declared Feb. 7 that the Soviet military personnel and equipment currently known to be in Cuba did not constitute a direct threat to the U.S. or other American nations.

Mr. Kennedy's assessment contradicted charges that these forces included combat units and possibly hidden ballistic missiles that directly threatened the western hemisphere.

The President challenged his critics to substantiate their charges. He warned that if the USSR had reintroduced aggressive missiles in Cuba, it would be responsible for the "greatest crisis the world has faced."

Mr. Kennedy said that the U.S. was pressing the "unfinished business" of the Soviet troops remaining in Cuba, and he hoped to obtain a commitment from Soviet Premier Khrushchev to carry out his pledge for their withdrawal.

Pres. Kennedy's statement on the Cuban military controversy, made at his Washington news conference, was one of a series of appearances made by Administration officials to refute charges that the Administration had concealed the extent and nature of Russian military activity in Cuba. The President's statement was preceded by detailed presentations in which Defense Secy. McNamara reported to a nation-wide TV audience and CIA Dir. McCone briefed Congress on the known facts of the Soviet military involvement in Cuba.° †

*McNamara, in a 2-hour TV-press interview at the State Department, said Soviet missiles had been removed from Cuba "beyond any reasonable doubt." Declaring that the U.S. would "not tolerate the use of Cuba as a base for the export of aggression, subversion," McNamara said the U.S. had "positive evidence" that Soviet missile submarine bases had not been established in the Caribbean. McNamara gave this U.S. "estimate of Soviet military strength in Cuba": Soviet personnel there numbered about 17,000, including technicians and 4 combat forces totaling about 5,000 men; the 17,000 Russians represented a drop of 5,000 from the estimated 22,000 in Cuba Nov. 1; the U.S. had evidence that "rocket equipment" of Soviet combat forces was being withdrawn; the U.S. was maintaining photographic surveillance of Cuba at least once a week, all Soviet ships arriving at and leaving Cuba were watched; about 102 MiG fighters, including MiG-21s were in Cuba but had no nuclear bombing capacity.

† McCone's denial of a Soviet military build-up in Cuba was made Feb. 6 in testimony before an Armed Services subcommittee, headed by Sen. John C. Stennis (D., Miss.), which opened secret hearings into the Cuban situation that day. Major points of a censored version of McCone's testimony: (1) U.S. intelligence agencies, having checked all rumors, were "convinced beyond reasonable doubt" that all Soviet missiles and bombers in Cuba "were withdrawn" soon after the U.S. naval quarantine ordered by Pres. Kennedy Oct. 24. (2) "There have been a number of reports that offensive weapons have been concealed in caves. Some of these reports evidently derive from the known Cuban practice of using caves for storage of small arms, ammunition and other [military] items. . . ."

Answering reporters' questions as to whether he regarded Cuba—and the Soviet forces on the island—as a serious military threat, the President said:

"I think we ought to keep a sense of proportion about the size of the [Soviet] force we are talking about. We are talking about 4 groups, 1,100 to 1,200 men each. Those are the organized military units. That is about 6,000 men. Obviously, those forces cannot be used to invade another country. They may be used to maintain some sort of control within Cuba, but, obviously, are not a force that can be used externally. And, in addition, Cuba . . . lacks amphibious equipment, and . . . our power in that area is overwhelming." "It may be that there are hidden away some missiles. Nobody can prove . . . that they are not there, or they might be brought in. But they are going to have to be erected, and we continue complete surveillance. They have to be moved. They have to be put onto pads. They have to be prepared to fire. . . . If the Soviet Union did that, it would indicate that they were prepared to take the chance of another great encounter between us, with all the dangers."

Mr. Kennedy's statement clearly expressed his doubt that the USSR was prepared to provoke a war crisis to attain its policy objectives in the Caribbean area.

Khrushchev Denies Reports. The charges of aggressive Russian military activities in Cuba were rejected by Soviet Premier Khrushchev Feb. 9 during a 2½-hour interview in the Kremlin with Roy H. Thomson, Canadian newspaper publisher.

Thomson told Western newsmen in Moscow that he had informed Khrushchev of the U.S.' concern over the Soviet troops in Cuba and had asked whether there was any basis for this concern. Khrushchev, he said, had replied: "None at all."

Khrushchev reportedly gave Thomson the following assurance that there were no Soviet nuclear weapons in Cuba: "All the atomic weapons [Khrushchev said] are in our hands, and we do not need to place them there. We can reach anyone we want with our own weapons from our own territory. It is more reliable to have them on our own territory."

According to Thomson, Khrushchev professed not to know the precise number of Soviet personnel in Cuba but asserted that the troops sent there were engaged in training, not combat duties. He asserted that "many [of the Soviet troops] have left Cuba and are continuing to leave Cuba."

Soviet Withdrawal Pledge. The USSR informed Pres. Kennedy Feb. 18 that "several thousand" of the Soviet Union's estimated 17,000 troops in Cuba would be withdrawn. A Soviet message delivered to the President by Amb.-to-U.S. Anatoly A. Dobrynin reportedly set Mar. 15 as the deadline for the departure of the Soviet soldiers.

The Soviet message was disclosed by Mr. Kennedy at a special White House meeting he called later Feb. 18 of Congressional leaders and top Administration officials.

The Soviet announcement of the planned troop withdrawal reportedly was not a new commitment by Moscow but rather a response to Mr. Kennedy's demands for clarification of a Nov. 20, 1962 note in which Premier Khrushchev had promised that Soviet troops in Cuba would be pulled out "in due course."

Mr. Kennedy reportedly told the White House meeting that clamor in the U.S. over removal of the Soviet troops made it more difficult for Khrushchev to withdraw them because of his domestic political situation.

Senate Republican leader Everett M. Dirksen (Ill.), who attended the

White House meeting, said Feb. 19 that "there was no indication in the Dobrynin message of how many troops, or what kind, will be withdrawn or when."

Chrmn. Stennis of the Senate Preparedness Subcommittee, who also had attended the meeting, said Feb. 19 that U.S. "policy must be an insistence that they all [the Soviet troops] be withdrawn" "to really remove the menace."

U.S. Warned Vs. Invasion. Soviet Premier Khrushchev warned Feb. 27 that if U.S. "imperialists" invaded Cuba, the USSR would come to the aid of the Castro government "and strike a devastating blow at the aggressors." Khrushchev's warning was made in a radio-TV election campaign speech in which he pledged similar Soviet assistance to Communist China, North Korea, North Vietnam, East Germany, "or indeed any Socialist country."

Khrushchev said: ". . . American imperialists have not apparently renounced the policy of aggression and provocations [against Cuba]. The most aggressive American circles, the 'wild men,' as they are called by the Americans, are urging the American government to invade Cuba on the pretext that she allegedly poses a threat to the United States . . . [and] claiming that one cannot tolerate a differently minded neighbor. . . . If the United States undertook aggressive actions against Cuba and the Soviet Union against its different-minded neighbors, chaos would develop which would inevitably lead to war."

The U.S. "cannot but understand that an invasion of Cuba . . . would be incompatible with the commitments assumed during the [1962] crisis in the area of the Caribbean. . . . [Pres. Kennedy] pledged himself not to invade Cuba while we agreed to withdraw strategic missiles and IL-28

bombers from Cuba, and did withdraw them. But this does not mean that we have left . . . Cuba at the mercy of the big shark of imperialism. We gave the Cuban people our word . . . that the Soviet Union will come to Cuba's aid, and we shall not leave her in the hour of need."

Departures Reported. According to reports appearing in the Soviet and Western press, about 3,700 to 3,800 Soviet troops left Cuba on 5 ships between Feb. 5 and Mar. 15. The greatest number of them, reportedly about 2,000, left Mar. 15 aboard the Soviet passenger liner *Admiral Najimov.*

The Soviet Communist Party newspaper *Pravda,* discussing the departures, said Mar. 13 that the Soviet soldiers were leaving in accordance with a USSR-Cuban armed forces agreement. *Pravda* said: The Russian troops had completed the task of helping "their Cuban comrades to master fully the powerful modern Soviet military techniques which the revolutionary armed forces of . . . Cuba now possessed, to protect their country from any intrigues of the imperialist aggressors"; the Soviet military specialists leaving the island had helped the Cubans "in the construction of defense fortifications and airfields where now are stationed airplanes of the Cuban Air Force."

Cuban Premier Castro asserted Mar. 13 that "the arms which we have and those which are passing into our hands are adequate to repel anything they [the U.S.] may send against us." Castro said his government was "recruiting personnel for those modern arms. We need technicians. We shall select them from the armed forces and from centers of labor."

Pres. Kennedy said at a news conference in Washington Apr. 3 that about 4,000 Soviet troops had left Cuba in March. These were in addition to the 5,000 that departed in Nov. 1962. Mr. Kennedy said this troop exit "still leaves some thousands [an estimated 12,000-13,000] on the island," and "we hope they are going to be withdrawn."

Rusk Vs. GOP Critics. State Secy. Rusk defended the Administration's Cuban policies—and particularly its position on the Soviet military presence in Cuba—in an address Apr. 18 at a Washington meeting of the American Society of Newspaper Editors. Rusk's statements were made in reply to a speech earlier that day in which Sen. Keating had reiterated his criticism of the handling of the Cuban situation.

Rusk declared that the U.S. was pursuing a "serious, sustained and effective" policy toward Cuba. It had, he said, made it clear to the USSR that the presence of Soviet combat troops in Cuba was unacceptable. Rusk said that the U.S. already had begun carrying out a 3-point program to contain the Cuban menace: (1) U.S. armed forces were maintaining a close air and sea surveillance of Cuba to prevent the re-entry of Soviet weapons; (2) the U.S. was making it clear that there was "no future for Cuba under the present régime" because the Cuban government was economically and politically isolated from the rest of the Western hemisphere; (3) the U.S. was halting the "clandestine flow of funds and personnel to and from Cuba" in an effort to check the export of subversion and communism to the rest of Latin America.

Sen. Keating, in his speech, had rejected Pres. Kennedy's Apr. 3 news conference statement that 4,000 Soviet troops had left Cuba in March and that about 13,000 were left on the island. Contending that Cuba currently had as many Soviet troops as in Nov. 1962, Keating said: "There were and are now in the neighborhood of 17,000 there"; "no functional units have been withdrawn"; the publicized departure of Soviet troops from Cuba

had been balanced by the unpublicized arrival of "a comparable number of new troops."

Ex-Vice Pres. Richard M. Nixon addressed the editors' meeting Apr. 20 and demanded that the U.S. take action against Cuba as part of a new and aggressive foreign policy designed to stem the spread of communism to new areas of the world. Nixon called for a U.S. "command decision to do whatever is necessary to force the removal of the Soviet beachhead" in Cuba. Specifying that the U.S. should use only the "minimum" force required to accomplish this task, he urged that the U.S. adopt a "freedom doctrine" for the Americas and give open support to all opponents of Communist régimes, including Cuban refugees.

(New York Gov. Nelson A. Rockefeller, in a New York speech Apr. 20, called the presence of Soviet troops in Cuba "the clearest evidence of failure" of the Administration's foreign policy. "We've got to do something about it," he said. He called it "hard" for Americans "not to support . . . people who want to regain their freedom.")

Report Doubts Red Threat. The Senate Preparedness Subcommittee investigated the alleged Soviet military build-up in Cuba and made public a censored version of its report May 9. Excerpts from the report:

▶ According to intelligence sources "approximately 5,000 Soviet personnel were withdrawn from Cuba, following the October confrontation, leaving . . . about 17,500 Soviets in Cuba. A net of 4,000 to 5,000 additional have been withdrawn since the first of the year. . . ."

▶ ". . . the combined Soviet and Cuban forces now in the island . . . are admittedly capable of suppressing any internal rebellion . . . and an invasion . . . would require large forces. . . ."

▶ ". . . the intelligence chiefs do not believe that the Communist forces in

Cuba now present a direct threat to the U.S. or Latin America."

▶ ". . . Castro is supporting . . . Communist revolutionary . . . movements throughout the Western Hemisphere, and . . . such activities present a . . . threat to the peace and security of the Americas."

JFK: Most Reds Instructors. Pres. Kennedy said at his news conference in Washington Aug. 20 that since June there had been "a decline" in the number of Soviet troops in Cuba. The President said "the primary emphasis of those [Soviet soldiers] who remain now is in training and not in concentrated military units." Asserting that it was "difficult" to estimate the number of troops withdrawn, Mr. Kennedy said. "On the information we have about outward movement and inward movement, it is the judgment of the intelligence community that there has been a reduction in the last 2½ months." The June estimate of Soviet troops in Cuba: 12,500, of which 4,-000-5,000 were believed members of combat units.

5,000 Said to Remain. The *N.Y. Times* reported Nov. 10 that U.S. government officials had estimated the number of Soviet troops in Cuba to be down to 5,000. The report contended that there were no longer any Russian combat units on the island. The estimates, based on U.S. aerial surveillance and other intelligence sources, indicated that Soviet troops had been leaving the island at the rate of 500 a week for the past 3 months. The remaining Soviet troops were said to be used to train Cuban soldiers and to maintain Soviet military equipment.

State Secy. Rusk said at a news conference Nov. 8 that there had been "substantial withdrawals" of Soviet personnel from Cuba, but he refused to specify the number remaining.

(However, the House Republican Policy Committee, apparently quoting Cuban exile sources, asserted Nov. 8 that departing Soviet troops from Cuba were being replaced by units from Communist China and by personnel from other Communist countries in East Europe. Commenting on this statement, the Defense Department said Nov. 8 that "there may be a few Chinese advisers, observers and visiting delegations on the island, [but] we have no evidence of any Chinese military units in Cuba.")

Hemisphere Action Against Cuba

The U.S. and its Latin American allies regarded Cuba as a subversive threat to the Western Hemisphere despite the apparent abandonment of efforts to turn the island into a Soviet base in the Caribbean. Continued Cuban efforts to instigate Latin revolts were charged by U.S. intelligence sources and the Organization of American States. In an effort to confine the Castro threat, the U.S. undertook measures to strengthen the economic blockade it had imposed on Cuba in 1962: it tightened foreign trade restrictions, urged its Western allies to halt shipments to Cuba, froze Cuban assets in the U.S. and rallied its Latin allies in the OAS and at a conference of Central American nations held in San José, Costa Rica.

Central American Conference. Pres. Kennedy and 6 Central American presidents met in San José, Costa Rica Mar. 18-19 to discuss the U.S.' Alliance for Progress, the economic integration of Central America and the Cuban threat to the existing Western Hemisphere system.

The 6 other presidents participating in the San José meetings were Francisco J. Orlich of Costa Rica, Julio A. Rivera of El Salvador, Miguel Ydígoras Fuentes of Guatemala, Ramón Villeda Morales of Honduras, Roberto F. Chiari of Panama, and Luis Somoza

DeBayle (and Pres.-elect René Schick Gutierrez) of Nicaragua.

Opening the conference Mar. 18, Pres. Kennedy pledged that the U.S. would provide assistance for Central America's economic development and would contain the Castro régime to permit other hemisphere nations to live and work in peace. Mr. Kennedy declared: "We will build a wall around Cuba—not a wall of mortar or brick or barbed wire, but a wall of dedicated men determined to protect their own freedom. . . ." The President reminded the conferees that while "newly independent nations rise in the Caribbean, the people of Cuba have been forcibly compelled to submit to a new imperialism, more ruthless . . . in its pursuit of power than any this hemisphere has known."

A Declaration of San José was signed at the conclusion of the conference Mar. 19. Its economic section contained a $6 million U.S. pledge for a regional survey to determine the timing and type of U.S. aid to be given Central America. The declaration's political section promised further efforts to halt Soviet aggression in the Western Hemisphere and expressed the conferees' agreement that communism was the principal obstacle to Central America's economic development.

Although Cuba was not specifically mentioned in the declaration, Mr. Kennedy, speaking at his Washington news conference Mar. 21, said that the San José meeting had "agreed on the necessity . . . to halt the flow of agents, money, arms and propaganda from Cuba to Central America."

U.S. Subversion Inquiry. The House Inter-American Affairs Subcommittee opened hearings Feb. 18 on Communist subversion in Latin America.

The subcommittee's first witness, Asst. State Secy. (for Inter-American Affairs) Edwin Martin, said Cuba had

embarked on a program of "open encouragement of terror as a principal weapon" in subverting other Latin American countries.

Another witness, CIA Dir. John A. McCone, charged Feb. 19 that Premier Castro was "spurring and supporting the efforts of Communists and other revolutionaries to overthrow and seize control of the governments of Latin America."

McCone, whose statement was released Mar. 1, said that according to U.S. intelligence information at least 1,000 to 1,500 Latin Americans had gone to Cuba in 1962 alone for training in guerrilla warfare and subversion. McCone said it appeared that Venezuela had been designated as "No. 1 on the priority list for revolution."

Cuban Shipping Curbed. Pres. Kennedy issued an order Feb. 26 barring transport of U.S. government-financed cargoes on foreign merchant ships that engaged in trade with Cuba. The ban covered material partly or completely financed by the U.S., including subsidized farm products and military and civilian goods shipped under U.S. foreign aid programs.

Pres. Kennedy's order followed reports of the continued participation of Western merchant ships in Communist bloc trade with Cuba. Among the Western vessels reportedly involved in such trade were those of Britain, Norway and Greece.

The State Department had warned all Western nations Jan. 11 that they risked losing U.S. assistance if they continued to permit their merchant ships to be involved in trade with Cuba. The warning was based on the U.S.' 1962 Foreign Aid Act, which denied U.S. assistance to nations whose ships engaged in such trade.

OAS Acts Vs. Subversion. The Council of the Organization of American States July 3 approved by a 14-1 vote a series of recommendations for action to combat alleged Cuban subversion in the Western Hemisphere. Chile voted against; 4 nations abstained: Brazil, Haiti, Mexico and Venezuela.

The recommendations, submitted by the OAS' 8-nation Inter-American Security Committee*, were based on an investigation which the committee had conducted at the request of the OAS Council. The inquiry, proposed by the U.S., had been approved by the Council Apr. 23 by a 13-1 vote (6 abstentions). (The U.S. resolution had called for the probe to be extended to any country without the approval of its government. Brazil, which had cast the only negative vote, favored the investigation but only in countries that asked the OAS to do so. These 6 nations had supported Brazil's argument and had abstained: Mexico, Venezuela, Chile, Bolivia, Haiti and the Dominican Republic.)

The committee's report suggested that American governments prohibit international pro-Communist meetings, bar the movement of funds and people to Cuba and establish ties among hemisphere security agencies.

The report said the committee's inquiry had resulted in the following findings:

▶ Soviet military equipment in Cuba had "converted" Cuba "into a Soviet military camp" and "a base for promoting subversive activities in other countries."

▶ The Castro régime practiced its subversion through culture centers in the Americas, through diplomatic and trade missions and through pro-Cuban conferences and travel to and from Cuba by agents with false papers and passports.

▶ Cuba had "begun a new phase of promoting . . . violence in other coun-

* Committee chairman—Juan Bautista Lavalle of Peru. Other members—Argentina, Colombia, Dominican Republic, El Salvador, Guatemala, Panama and U.S.

tries of this hemisphere." Venezuela had been selected as its "primary objective."

U.S. Freezes Assets. The U.S. July 8 froze $33 million in public and private Cuban bank deposits in the U.S. The U.S. also prohibited the unlicensed transfer of U.S. dollars to or from Cuba and barred all other unlicensed transactions with Cuba and all property dealings that involved a Cuban interest.

The State Department said the U.S. financial restrictions were imposed in compliance with the OAS Council's July 3 resolution urging Western Hemisphere nations to combat Cuban subversion.

Cuba retaliated against the U.S. move with a decree July 24 expropriating the U.S. embassy building in Havana. The building was occupied by Swiss embassy officials who represented U.S. interests in Cuba. The Swiss rejected Cuban demands to evacuate the building and the Castro régime failed to seize it.

Students Vs. Cuba Travel Ban. A group of U.S. students, most of them pro-Castro, who had defied the U.S. ban on travel to Cuba became embroiled in demonstrations at a Congressional hearing in Washington Sept. 12-13.

59 U.S. students, their travel expenses defrayed by Premier Castro, had arrived in Cuba June 30 in defiance of the ban. They flew to Havana from Prague after visiting London, Paris and Amsterdam. 50 of them returned to the U.S. by way of Madrid Aug. 29. (One of the group had returned to the U.S. Aug. 26, one died in Cuba, 3 stayed in Cuba, and 4 remained in Madrid.) On their arrival in the U.S. they became involved in a controversy with the State Department over the legality of their trip.

10 of the group were subpoenaed to testify at an investigation of the

Cuban Premier Castro joins Soviet Premier Khrushchev and other Soviet officials atop Lenin's Mausoleum in Red Square for May Day review of military and other marchers. Left to right: Emilio Aragones Navarro (chief of Cuban militia), Soviet Defense Min. Rodion Y. Malinovsky, Castro, Khrushchev, Soviet Pres. Leonid Brezhnev and Soviet First Deputy Premier Anastas Mikoyan.
(Tass photo from Wide World)

trip by the House Un-American Activities Committee under the chairmanship of Rep. Edwin E. Willis (D., La.). The hearings were recessed in disorder after the students and their sympathizers heckled the committee and called Southern committee members "racists." Police, called in to quiet unruly spectators, ejected several of them Sept. 12 and Sept. 13.

Members of the "student" group who testified Sept. 12: Barry Hoffman, 26, of Brookline, Mass., who described himself as a voluntary, "undercover agent." Lee Levi Laub, 24, of New York, who invoked the 5th Amendment in refusing to answer questions; Laub, a Negro, called committee members "racists."

Laub and Phillip A. Luce, 26, of New York were indicted by a federal grand jury in New York Sept. 27 on charges of conspiring to organize and promote the Cuba trip and of taking it in violation of U.S. travel laws.

Cuban-Soviet Relations

Castro Visits USSR. Premier Castro began a month-long visit to the Soviet Union Apr. 27. He returned to Havana June 3 after receiving renewed pledges of Soviet aid against military attack and, reportedly, assur-

ances that arrangements would be made to free Cuban sugar for sale at rising world prices.

Arriving at an airport near Murmansk, Castro was greeted by a Soviet delegation led by First Deputy Premier Anastas I. Mikoyan. After reviewing units of the USSR's Northern Fleet from a Soviet destroyer off Murmansk Apr. 28, Castro and his party were flown to Moscow, where they were greeted later that day by Premier Khrushchev. The 2 leaders drove together from the airport to Red Square, where they addressed a crowd of 40,000.

In his speech, Castro said: "Comrade Khrushchev today expressed his confidence in the victory of the Cuban revolution, and we are confident that will be so"; "were it not for the Soviet Union, the imperialists would not hesitate to launch a direct military attack on our country"; "it precisely was the might of the Soviet Union and the entire Socialist camp that deterred an imperialist aggression against our homeland"; Soviet arms and economic aid had prevented the destruction of the "Socialist revolution in Cuba."

Castro and his advisers began secret talks with Khrushchev and other Soviet leaders in the Kremlin Apr. 29. The Cuban leader attended a May

Day review in Red Square May 1, standing atop the Lenin Mausoleum with Khrushchev and other high Soviet officials. (U.S. Amb. Foy D. Kohler boycotted the celebration because of Castro's presence.) Castro interrupted his Moscow talks for visits to major cities in European and Siberian Russia May 7-17. He returned to Moscow and was the guest at a massive friendship meeting held at the city's Lenin Stadium May 23. In a speech at the latter event, Castro declared that the Cuban revolution had "irrefutably confirmed the Marxist-Leninist thesis that in the modern world the balance of forces is no longer in favor of the imperialist camps." Khrushchev also addressed the stadium rally, pledging again that the USSR would "discharge our duty . . . to the fraternal Cuban people and come to their assistance" if they were attacked by the U.S.

The 2 leaders issued a joint communiqué May 24. Its principle points: Khrushchev had accepted Castro's invitation to visit Cuba; in the event of an attack "in violation of the commitments undertaken by the United States President not to invade Cuba," the USSR would provide "the necessary help for the defense of . . . Cuba with all the means at its disposal"; Cuba had accepted a Moscow proposal to change the current Soviet-Cuban sugar agreement and to increase the price of Cuban raw sugar bought in 1963 to the world price level as "an effective rôle in strengthening the economic situation" in Cuba. (Further light was shed on the Cuban-Soviet sugar statement by the *N.Y. Times* June 2. The newspaper said that under an agreement negotiated in Moscow in Dec. 1962, the Soviet Union had permitted Cuba to divert to the world market in 1963 1,000,000 tons of sugar to take advantage of rising prices. The 1,000,000 tons was to come from Cuba's antici-

pated 1963 crop of 3½ million tons, all of which previously had been contracted to be sold to the USSR and Communist bloc nations.)

(Soviet First Deputy Foreign Trade Min. Sergei A. Borisov had reported Jan. 16 that 1962 Cuban-Soviet trade had totaled more than 500 million rubles [$550 million]. According to official Soviet publications, the 1962 Cuban-USSR trade goal had been $750 million.)

Castro Criticizes Khrushchev. Cuban Premier Castro had criticized Premier Khrushchev's stand in the Cuban missile crisis in an interview published by the Paris newspaper *Le Monde* Mar. 21-22. (Castro denied Mar. 22 that he had given the newspaper an exclusive interview. But he acknowledged he had held "informal talks" in Havana in January with Claude Julien, assistant foreign editor of *Le Monde.*)

In *Le Monde's* Mar. 21 installment, Castro was quoted as saying that Khrushchev "should not have removed his missiles without consulting us. Cuba does not want to be a pawn on the world's chessboard. Cuban sovereignty is a reality. I cannot agree with Khrushchev promising Kennedy to pull out his rockets without the slightest regard to the indispensable approval of the Cuban government."

Other points made by Castro:

▶ In offering rockets to Cuba, the USSR "explained to us that in accepting them we should strengthen the Socialist camp in the world. . . . That is why we accepted them. It was not to insure our own defense. . . ."

▶ The reasons for the placement of Soviet rockets in Cuba "is a mystery. Perhaps historians will succeed in clarifying that in 20 to 30 years."

▶ "Obviously the Soviet Union has world responsibility we do not have. Khrushchev wants peace, and we also want peace. He avoided war, but he did not win the peace."

▶ "The Chinese [Communists] are right in saying one should not yield to imperialism. But we . . . know that imperialism is not a paper tiger."

Exile & Underground Attacks

Anti-Castro commando groups continued their attacks on the Cuban revolutionary régime from within and outside the island. Exile groups, however, increasingly were frustrated in efforts to launch raids from the U.S. as the result of a Kennedy Administration decision to deny them havens for their operations. The decision was based on the belief that such raids were ineffective and on the fear that the attacks, which had been extended to Soviet ships in Cuban waters, might embroil the U.S. in hostilities beyond its control. The U.S. crackdown, however, did not completely thwart the raiders. Clandestinely using Caribbean islands as bases, some of them in the British Bahamas, exile groups continued their hit-and-run attacks on Cuba, employing light planes as well as boats.

U.S. Acts Vs. Exiles. The U.S. government announced Mar. 30 that it would "take every step necessary" to make sure that Cuban exile attacks "against Soviet ships and other targets in Cuba" "are not launched, manned or equipped on U.S. territory."

The U.S. position, as outlined in statements issued by the State and Justice Departments: Cuban exile attacks were "neither supported nor condoned" by the U.S. Pres. Kennedy "has pointed out that they may have the effects opposite to those presumably intended by those who carry them out; that is, they may strengthen the Soviet position in Cuba rather than weaken it. . . ." Although the U.S. sympathized with the Cuban exiles, the U.S. was not "prepared to see our own laws violated with impunity or to tolerate activities which

might provoke armed reprisals, the brunt of which would be borne by the armed forces of the U.S."

Pres. Kennedy had expressed opposition to the exiles' raids at his news conference in Washington Mar. 21. He said: "These kinds of raids" "strengthen the Russian position in Cuba and the Communist control of Cuba and justify repressive measures within Cuba, which might otherwise not be regarded as essential"; he did not think such raids increased "the chances of freeing Cuba."

Reiterating his opposition to the attacks, Mr. Kennedy said at his news conference Apr. 3 that the raids, if continued, "will bring reprisals, possibly on American ships. We will then be expected to take a military action to protect our ships. It may bring counteraction. I think that when these issues of war and peace hang in the balance that the United States government and authorities should . . . have a position of some control on this matter."

In a further move against the exiles, the U.S. was reported Apr. 30 to have ended financial aid to the Cuban Revolutionary Council, the largest and most influential of the 300 Cuban exile groups registered in Miami. The council, formed prior to the 1961 Cuban invasion, reportedly had received $100,000-$200,000 in aid from the U.S. each month. An official U.S. source said that "in the light of recent developments," the U.S. government "considers it inadvisable to continue to support the council."

Exile Raids. The anti-Castro exiles began their 1963 attacks against Cuba with a bold raid Mar. 18 on a Soviet army camp and the Russian freighter *Lvov* at the north coast port of Isabela de Sagua.

The raid was carried out by "commando units" of Alpha 66 and the 2d National Front of Escambray. Officials of the 2 groups gave this account of the raid at a Washington news conference Mar. 19: 2 speedboats, carrying exiles armed with machine guns and cannon, exchanged gunfire with Cuban Coast Guardsmen and Soviet infantry; the raiders then attacked another part of the port and were met by fire from the anchored *Lvov;* the Soviet ship was seriously damaged by the raiders' attack; about 12 Russians were wounded in the engagement.

The Cuban Armed Forces Ministry charged Mar. 20 that the attack on the *Lvov* had been "launched from the U.S. and Puerto Rican territory."

Another Soviet ship, the *Baku*, was attacked Mar. 26 in the Cuban port of Caibarien and reportedly suffered heavy damage. Commando L, a dissident offshoot of Alpha 66, took credit for the raid.

The USSR protested to the U.S. Mar. 27 and 29 over the *Lvov* and *Baku* attacks. In the Mar. 27 protest, the Soviet Union charged that the U.S. was "offering Cuban counter-revolutionaries its territories and material needs" and was thus responsible for a "dangerous aggravation of the situation in the . . . Caribbean . . . and throughout the world."

The exiles avoided further attacks on Soviet shipping, but were unremitting in their raids against Cuban targets. These were the major attacks reported:

▶ An exile group called the Cuban Freedom Fighters announced in Miami Apr. 26 that its 2-engine plane had attacked a Havana oil refinery the previous day. 4 Americans were on the plane, said to have taken off from a Caribbean island. One of the Americans, identified as free-lance photographer Alexander I. Rorke Jr. of New York, reported that the bombs dropped on the refinery had exploded. The Cuban government confirmed the raid Apr. 26 but denied the dropped bombs had exploded.

▶ Cuba reported May 19 that its shore guns at Tarara Beach, 10 miles from Havana, had forced a Cuban exile raider to flee after it had machine-gunned the beach. A Miami exile report said the attack had been carried out by an underground group in Cuba, not by an exile raider.

▶ An unidentified light plane bombed and strafed a sugar mill Aug. 15 in the Camagüey Province town of Moron, about 200 miles east of Havana. A 2d light plane attacked oil storage tanks Aug. 18 at Casilda, on Las Villas Province's south coast. The raid set one 8,000-gallon tank ablaze and punctured another.

▶ According to Cuban government reports, 2 jet bombers, escorted by fighters, attempted to attack an air force base at Santa Clara Sept. 5 but were driven away by Cuban anti-aircraft fire. The fleeing planes dropped their bombs around Santa Clara, killing one person and wounding 3 others. Raiding planes made Santa Clara their target for the 2d time Sept. 8 but were again driven off by anti-aircraft fire.

▶ Miami exiles claimed Sept. 8 that one of their planes had bombed the Brazil (formerly Jaronu) sugar mill, Cuba's largest, and eliminated it as a producer.

(The U.S. was reported to have received a 2d Soviet warning about exile attacks on Cuba Sept. 9. A Soviet Spanish-language broadcast beamed to Cuba said the USSR had told the U.S. that Moscow "will not tolerate" further raids on Cuba by anti-Castro exiles "armed and supplied with North American weapons." The U.S. State Department denied Sept. 10 that it had received such a warning from the USSR.)

The capture of several U.S. CIA agents engaged in a raid on Cuba was claimed by Premier Castro Oct. 30. Castro said the men had been seized aboard 2 small boats from a CIA-operated raider ship called the *Rex*, a 150-foot vessel flying the Nicaraguan flag. A 174-foot vessel named the *Rex*, flying the Nicaraguan flag, had returned to Palm Beach, Fla. Oct. 28 with 2 of its motor launches missing. Miami oil company executive J. A. Belcher acknowledged ownership of the *Rex* but denied it had participated in raids on Cuba.

With the mounting U.S. restrictions on exile raiders' operations from the American mainland, many of the anti-Castro organizations shifted their commando bases to the small Caribbean islands off Cuba. Several exile raids were launched from remote islands of the British-controlled Bahamas chain, and these operations, matched by an incident in which Cuban forces landed and seized a group of refugees that had fled to one of the islands, embroiled Britain in the controversy over exile activities.

Cuba reported Feb. 22 that it had captured a hijacked Cuban fishing vessel Feb. 21 that had left the Bahaman island of Elbow Cay, where it allegedly had picked up men and arms supplied by the CIA. Cuban navy ships intercepted the boat and its 8 occupants, who were said to be planning to arm the anti-Castro underground in Cuba.

British marines and Bahaman policemen Mar. 31 captured 16 Cuban exiles and one American (Jerry Buchanan of Miami) and their armed motor launch, *Violynn III*, on Norman Key in the Exuma Islands. The capture was based on information from U.S. authorities that the men were seeking to refuel their boat on the island in preparation for an attack on a Soviet tanker. Buchanan was released by British authorities with the 16 other crewmen Apr. 4. Alexander I. Rorke Jr. identified himself Apr. 2 as the owner of the *Violynn III*. He said the vessel had carried out 11 anti-Castro missions since Oct. 1962.

The Bahama islet of Anguilla Cay was the scene of an incident Aug. 13 involving Cuban refugees, Cuban armed forces and the U.S. The State Department reported Aug. 15 that a Cuban air and naval force had landed in Anguilla Cay Aug. 13 and seized 19 of a group of 29 Cuban refugees who had fled there from Matanzas Province that day in a hijacked boat. The U.S. Coast Guard evacuated the 10 remaining refugees to Miami, at British request. A Cuban government version of the incident said the fleeing refugees had been captured aboard 2 stolen fishing vessels near Anguilla Cay, but it made no mention of a landing on the island. A leader of the refugee group said that a Cuban helicopter machinegunned the exiles on the beach and that a Cuban landing force then had arrived and captured 19 escapees.

U.S. Coast Guardsmen and airmen had witnessed the seizure of the refugees, but State Secy. Rusk said Aug. 16 that the U.S. force could not have intervened because it did not have British authority to do so.

Cuba rejected a British demand Aug. 22 that Havana apologize for the raid and return the 19 refugees.

Underground Attacks. Actions carried out by the anti-Castro underground in Cuba were reported both by the Castro government and Cuban exile sources outside the country.

Among the most important reports of underground activities: Havana announced Jan. 27 that it had crushed 2 U.S. "spy networks" in Oriente Province. The purpose of the alleged networks, according to a government communiqué: assassination of Armed Forces Min. Raúl Castro and the precipitation of revolts "to protect a massive landing of counter-revolutionaries" in February. The government statement said "a large number of agents" had been arrested and arms and explosives had been seized.

Cuban militiamen fired at 4 anti-Castro Cubans Mar. 14 as they tried to drive a stolen army jeep through a wire fence surrounding the Uruguay embassy in Havana. One of the men was killed, a 2d was wounded and the 2 others were captured.

The Cuban government announced May 15 the smashing of a "counter-revolutionary" band that had carried out acts of sabotage and terror in the Matanzas Province town of Guira de Melena, 40 miles south of Havana. The announcement said several of the rebels were killed. Matanzas unrest was further acknowledged by Premier Castro, who said in a speech June 11 that his forces had wiped out 14 rebel groups in the province.

Cuban firing squads executed 4 persons July 31 for counter-revolutionary activities. The Cuban radio said 3 others had received 20-year jail terms and a 4th was given a 12-year sentence. The government announced the execution of 3 more anti-Castro Cubans Aug. 16 on charges of having landed in Cuba under CIA orders to organize military espionage and sabotage. The announcement said the 3 had been captured July 22 in Cayo Verde off the north coast of Las Villas Province.

Exile sources in Miami Sept. 10 reported that widespread clashes had occurred between anti-government guerrillas and Cuban militiamen in Oriente Province and in the Sierra de Escambray area. A secret rebel radio in Oriente was quoted as reporting that a guerrilla force had captured a militia camp Sept. 5 at Arroyo Blando and had killed 64 militiamen; 6 guerrillas were slain.

Another Miami exile report Sept 10 told of a clash a few days previously between 1,500 militiamen and an undisclosed number of rebels near Guasimal; 80 militiamen and 18 guerrillas were said to have been killed. The report said that after the battle

authorities arrested about 1,500 residents of Guasimal and 3 nearby towns.

U.S. Boats Attacked. The U.S. State Department reported Feb. 21 that 2 Cuban MiGs had attacked a disabled U.S. shrimp boat, the *Ala*, Feb. 20, 60 miles north of Cuba. The boat's 2 occupants were not injured and were picked up by a U.S. destroyer and brought to Key West, Fla.

A U.S. protest to Havana Feb. 21 called the attack "unprovoked and willfull."

The Defense Department received a Presidential order Feb. 21 to "take all necessary" action to prevent other similar attacks.

Cuba Feb. 22 "rejected and denied" the U.S. charge that Cuban planes had attacked the *Ala*. Havana said the MiGs only "had flown low" over the *Ala*.

2 Cuban MiGs Mar. 28 fired on but missed the 5,000-ton U.S. motorship *Floridian* about 25 miles off Key West. The ship returned to Miami Mar. 29 under U.S. military plane escort. Cuba Mar. 29 admitted and apologized for the attack.

The 31,505-ton U.S.-owned ore ship *J. Louis* was attacked in apparent error by Cuban MiGs Oct. 22 as it sailed into the area of a clash between Cuban forces and a raiding party of Cuban exiles 13 miles off the southeastern tip of Cuba. The Liberian-registered ship, en route from Jamaica to Corpus Christi, Tex., radioed that a jet plane of "unknown identity" had made attacking passes over it for more than an hour and set it afire. The blaze was extinguished in 3 hours, and the *J. Louis* resumed its journey; it arrived in Corpus Christi Oct. 23.

Dispute Over Exile Invasion

The controversy stirred up by the ill-fated invasion of Cuba in 1961 continued to make headlines 2 years later. The dispute centered on 2 ques-

tions: Did the U.S. withdraw an alleged promise to provide air cover for the landings at the Bay of Pigs? Did the Kennedy Administration promise the Cuban exiles that it would support another invasion of Cuba after the first one had ended in disaster? The differing answers given to these questions created controversy on the U.S. political scene and exacerbated relations between the U.S. and Cuban exiles.

One aspect of the Cuban invasion was successfully concluded: The exchange of refugees in Cuba for U.S. goods, promised in the 1962 agreement for the ransom of the Bay of Pigs prisoners.

JFK Denies Air Cover Pledge. Recurrent reports that the U.S. had pledged air cover for the 1961 rebel invasion of Cuba but at the last minute had decided not to do so were denied by Pres. Kennedy at his Washington news conference Jan. 24.

Mr. Kennedy said that such air cover would have involved the commitment of U.S. forces to a full-scale invasion, which "was not the policy of the United States in 1961." The President said the only U.S. air cover planned or provided in the invasion was made up of World War II B-26 bombers flown by Cuban refugees from bases outside the U.S. The B-26s were ineffective because the Cuban air forces' T-33 jet trainers, formerly U.S. planes, were able to maintain air supremacy over the Bay of Pigs invasion beach, Mr. Kennedy said. The President acknowledged again that the White House was responsible for the failure of the invasion.

Atty. Gen. Robert F. Kennedy had made similar denials of U.S. promises for air cover in a Knight Newspapers interview published Jan. 21. Kennedy discounted charges that the invasion had failed because the President, at the last moment, had withdrawn a pledge to provide U.S. air power to support the invaders. He confirmed that the invaders had appealed for U.S. naval support after they were pinned down on the beach at the Bay of Pigs. The U.S., he said, had rejected the plea and had adhered to "the ground rules . . . accepted by everyone that there would be no U.S. forces in the invasion under any condition."

Other points made by the Attorney General:

▶ "The President has taken responsibility for the failure. . . . He approved the plan that was fully cleared by the CIA and the Joint Chiefs of Staff. It was 'wargamed' at the Pentagon. . . ."

▶ "There was not sufficient air cover at the beach. . . . There were not enough men and equipment. . . . The planning was inadequate."

▶ The principal purpose of the invasion was to establish a beachhead to enable the Cuban invaders to set up a Cuban government-in-exile that would eventually be recognized by the U.S.

Sen. Barry M. Goldwater (R., Ariz.) contradicted the Administration version of events surrounding the 1961 invasion. In an interview Jan. 21, Goldwater said that he had conferred with Pres. Kennedy after the invasion and "got the impression that an air cover had been part of the original invasion plans." "I am sure the entire American public has understood that the air cover was definitely in the invasion plans until the President was persuaded—by some still unidentified advisers—to cancel it."

Goldwater reiterated his charges Jan. 23 in a statement in which he said: "There was no denial from the White House nor from the Attorney General when stories were printed throughout the American press that air cover was withheld on orders of the President."

Miró Charges Attack Pledged. Conflicting versions of alleged U.S. promises to support another invasion of Cuba were voiced in April by Pres. Kennedy and by José Miró Cardona, president of the Miami-based anti-Castro Cuban Revolutionary Council. Speaking at a council meeting Apr. 18, Miró accused Mr. Kennedy of "breaking promises and agreements" for a new invasion of Cuba in favor of "peaceful coexistence" with the Castro régime. Miró made the accusation in a 10,000-word document he submitted to the council along with his formal resignation as the organization's president. (8 other council officials also resigned. Antonio Maceo was named as Miró's successor, but he quit the post June 22 in protest against what he regarded as the council's indiscreet disclosure of an exile commando raid on Cuba June 20. The council elected Vice Pres. Manuel Antonio de Varona as president June 24.)

Recapitulating his dealings with Pres. Kennedy following the failure of the 1961 Cuban invasion, Miró said: In a "pact" concluded by Mr. Kennedy April 20, 1961 (the day the invasion had ended in failure), "the Cubans were assured of a place in the vanguard" of a new "armed attack."

Miró said that at a meeting with the President May 4, 1961 Mr. Kennedy "planned with me the immediate future of Cuba," and "his offers of cooperation were definite and his backing was total and absolute." Miró added: As a result of the May 4 meeting, U.S. "support was made possible for the underground forces in Cuba, and the first program of recruitment of Cuban volunteers" in U.S. armed forces units was planned.

Atty. Gen. Robert F. Kennedy arranged a meeting for Miró with Pres. Kennedy April 10, 1962. At the meeting, Miró said, the President "assured me that the 'problem is essentially military and requiring 6 divisions,' that the [Revolutionary] Council should contribute the major contingent of soldiers." Miró said he had left the White House meeting "with the certainty that liberation was nearing for our country, with Cubans in the vanguard of combat."

But Miró added that he had become completely disillusioned as a result of the U.S. decision to halt Cuban exile attacks on Cuban soil and on Soviet shipping in the Cuban area.

Miró's charges precipitated disagreement in exile ranks. A statement issued by the Authentic Cuban Revolutionary Party, headed by Dr. Sergio Megias, declared that the Revolutionary Council did "not represent the will" of those fighting Castro. De Varona said Apr. 19 that he "never knew of a promise by Pres. Kennedy for a 2d invasion of Cuba."

President Denies Commitment. Pres. Kennedy asserted Apr. 19 that "quite obviously" no one in his Administration had promised another invasion of Cuba. The President's assertion was made at an impromptu news conference at a Washington meeting of the American Society of Newspaper Editors.

Reaffirming a State Department denial of Miró Cardona's accusations, Mr. Kennedy said: "We have not determined that it's in our national interest or in the general interest of the hemisphere for us to launch an invasion. And naturally that disappoints the [Cuban] exiles." "The foreign policy of the U.S., when so much depends upon us, must be made by the U.S." Despite U.S. sympathies for the Cuban exiles, the U.S. "cannot launch itself into a massive invasion of Cuba without considering the worldwide implications to other free countries and also its effect upon our own position."

Mr. Kennedy denied that the U.S.

government had "ever informed . . . Dr. [Miró] Cardona or anyone else, that we were going to launch, or committed ourselves to launch, a military invasion with 6 divisions."[*]

Refugee Exchange Ends. The U.S.-Cuban arrangement to exchange Cuban refugees for $53 million worth of U.S. food and drugs ended July 3 as the last shipment of refugees, 1,204, arrived aboard a freighter at Port Everglades, Fla.

The American Red Cross reported in Washington that 9,703 persons had been brought out of Cuba during the exchange, begun in Dec. 1962 with the release by Cuba of 1,113 prisoners taken during the 1961 exile invasion. James B. Donovan, New York attorney, had negotiated the release of the prisoners and refugees in conferences with Premier Castro in 1962.

(Dr. Manuel Urrutia Lleo, 62, Cuba's first president after the Castro revolution, was allowed Mar. 25 to leave the Mexican embassy in Havana, where he had been a political refugee since Nov. 1961. Urrutia arrived in Miami Mar. 29 to take up permanent residence in the U.S. Castro had removed Urrutia as president July 17, 1959 after Urrutia publicly expressed opposition to communism. Urrutia was given political asylum in the Venezuelan embassy in Havana but transferred to the Mexican embassy after the Caracas government ended diplomatic relations with Cuba.)

Castro Scores Exchanges. Premier Castro July 26 accused "the American government of not complying with" the prisoner-refugee agreement. The U.S., Castro said, "owes us $10 million" to complete payment of the prisoner-refugee ransom. Castro declared that the original agreement (approved but not sponsored by the U.S.) had called for the delivery to Cuba of $53 million worth of supplies but that only $43 million worth had been received.

The U.S. State Department noted July 27 that the U.S. had been informed of "Cuban Red Cross approval with the manner in which the agreement had been carried out."

[*] The President told a Washington news conference Sept. 12 that it was not "in the interest" of the U.S. to launch a "military invasion of Cuba." Mr. Kennedy said such an invasion would be "a most dangerous action, . . . which could bring a good deal of grief not only on the people of the U.S. but on Western Europe and others who are dependent upon us."

INDEX

Index entries refer to text by page number usually followed by the letter L or R to indicate whether the item appears in the Left or Right column

A

ADMIRAL Najimov (Soviet liner)—119R
ADZHUBEI, Aleksei I.—99L
AERIAL Reconnaissance—114L-116L, 121L
AGRICULTURE, Department of (U.S.)—6R, 10R, 11R, 12L, 30L
AGUIRRE, Severo—111L
AGUIRRE Fidaulet, Gen. Francisco—8L
AID (U.S.)—17R, 18R; see also ALLIANCE for Progress
AIKEN, Sen. George D. (R., Vt.)—116R
ALA (U.S. shrimp boat)—129R
ALBANIA—96, 99L
ALBIZU Campos, Pedro—47R
ALGERIA—19L, 110
ALLIANCE for Progress—60R-62L, 100R, 121R, 122L
ALMEIDA, Maj. Juan—25L, 111L
ALONSO, Luis—25L
ALPHA 66—107R, 126R-127L
ALSOP, Stewart—90L
ALVAREZ, Arturo—26L
ALVAREZ Castallón, Ceferino—46R
ALVAREZ Diaz, José—36R
ALVAREZ Román, Luis—111L
AMADEO, Julio—101L
AMADOR Quesada, Braulio—113L
AMERICAN & Foreign Power Co. of New York—12L
AMERICAN Legion—31R
AMERICAN Society of Newspaper Editors—120L-R
AMERICAN States, Organization of—see ORGANIZATION of American States
ANDERSON Jr., George W.—86L
ANDERSON, Howard F.—51L
ANDERSON, Robert B.—108R
ANDERSON Jr., Maj. Rudolf—79
ANFUSO, Rep. Victor L. (D., N.Y.)—45L
ANDUIZA, Mrs. Piedad—52R
ANSONOV (Soviet freighter)—82R-83L
ARAGONES Navarro, Emilio—110R, 112R, 124
ARBENZ Guzmán, Jacobo—31L
ARCAYA, Ignacio Luis—17R
ARGENTINA—14L, 28L, 35R, 40L, 69L, 100L, 101L, 123L
ARTIME, Capt. Manuel—36R, 37R, 39R, 47R, 108R
ATUNA, Vicentina—53R
AVIATION: Planes 'Hijacked'—56R-58L
AVINO Soler, Andres—52R-53L
AZORES—22L

B

BAHAMAS—128L-R
BAINER, Roy—47R
BAKER, Tommy L.—50L
BAKU (Soviet freighter)—127L
BALL, George W.—90R
BANKS & Banking—24L, 123R
BARALT, Luis A.—27L
BARRETO, Mrs. Berza—108R

BARRY, U.S.S. (destroyer)—82R
BARTLETT, Charles—90L-90R
BAYO, Gen. Alberto—58L
BAY of Pigs—37-44
BEANE, James B.—50L
BEARDEN, Leon & Cody—57
BEATON, Capt. Manolo—25L
BECK, George R.—50L
BELCHER, J. A.—128L
BELGIUM—6R-7L, 22L
Ben BELLA, Ahmed—110L-110R
BERLE Jr., Adolf A.—60L
BERMAN, Maximo—111L
BETANCOURT, Rómulo—59R, 110L
BETANCOURT, Mrs. Virginia—108R
BISHOPP, Earle N.—109R
BISSELL Jr., Richard M.—38R, 43L
BLOCKADE—65R, 66R-68L, 70R-72L, 89L, 90-91, 94L; see also EMBARGO
BOHEMIA Libre (Havana magazine)—105R
BOLAND, Frederick H.—24R
BONSAL, Philip W.—5-6, 9L, 22R
BORISOV, Sergei A.—125R
BORSINGER, Melchior—84L
BOSSI, Walter—33R
BOWLES, Chester W.—90R
BOZA Masvidal, Bishop Eduardo—29R, 51R-52L, 113R
BRADDOCK, Daniel M.—7, 9R, 22R, 26L
BRADLEY, Leslie—9L
BRAVO, Flavio—111L
BRAVO Puig, Leonardo—52R
BRAZIL—17R, 28L, 60L, 69, 100L, 101L, 123L, 127R
BREZHNEV, Leonid—99L, 124
BROADCASTING—28R, 29R
BRODERICK, Vincent L.—109R
BRUCE, Rep. Donald C. (R., Ind.)—117L
BRUNET, Eustace Dan-50R
BUCH Rodriguez, Luis—111L
BUCHANAN, Jerry—128L
BULGARIA—56L, 95L
BUNDY, McGeorge—90R

C

CABELL, Gen. Charles Pearre—6R
CABRERA, Caridad—26R
CADON, Albert Charles—57R
CAICEDO Castilla, José Joaquín—100R
CALCINES, Ramón—111L
CALIXTO López & Co.—30L
CAMPOS, Miguel—52R
CANADA—24L, 27L, 31L, 102, 103R
CAPEHART, Sen. Homer E(arl) (R., Ind.)—47L, 90L, 104R-105L
CARBO, Ulises—46R
CARCANO, Miguel Angel—101L
CARILLO, Francisco—45R
CARILLO, Santiago—58R
CARILLO Hernández, Dr. Justo—36R
CARRERAS Zayas, Maj. Jesús—50R
CARSWELL, Daniel L.—50L-R
CASIELLES Amigo, Julio—50R
CASUO Pérez, Enrique Gustavo—51L

133

134   Index